D0438363

FUZZY SYSTEMS THEORY and ITS APPLICATIONS

Edited By

Toshiro Terano
Faculty of Engineering
Hosei University

Kiyoji Asai
Faculty of Engineering
Osaka Institute of Technology

Michio Sugeno
The Graduate School of Nagatsuda
Tokyo Institute of Technology

Translated By
Charles G. Aschmann, III

ACADEMIC PRESS, INC.
Harcourt Brace & Company, Publishers

Boston   San Diego   New York
London   Sydney   Tokyo   Toronto

This book is printed on acid-free paper. ⊗

Original Japanese edition published as "Fajii Shisutemu Nyumon"
edited by Toshiro Terano, Kiyoji Asai and Michio Sugeno
Copyright © 1987 by Toshiro Terano, Kiyoji Asai and Michio Sugeno,
First published in 1987 by Ohmsha, Ltd., Tokyo, Japan
English edition copyright © 1992 by Academic Press
English translation rights arranged with Ohmsha, Ltd.

All rights reserved.
No part of this publication may be reproduced or
transmitted in any form or by any means, electronic
or mechanical, including photocopy, recording, or
any information storage and retrieval system, without
permission in writing from the publisher.

ACADEMIC PRESS, INC.
1250 Sixth Avenue, San Diego, CA 92101-4311

United Kingdom Edition published by
ACADEMIC PRESS LIMITED
24–28 Oval Road, London NWl 7DX

Library of Congress Cataloging-in-Publication Data

Terano, Toshirō, date.
    [Fajii shisutemu nyūmon. English]
    Fuzzy systems theory and its applications / Toshiro Terano, Kiyoji
Asai, Michio Sugeno.
        p.   cm.
    Translation of: Fajii shisutemu nyūmon.
    Includes bibliographical references.
    ISBN 0-12-685245-6
    1. Fuzzy systems.   2. Fuzzy sets.   I. Asai, Kiyoji.
II. Sugeno, Michio.   III. Title.
    QA248, T4213   1991
511.3'22—dc20   90-21586
                                                           CIP

Printed in the United States of America
93 94 95 96 97   EB 9 8 7 6 5 4 3

511.322
F996
1992

CONTENTS

Mills College Library
Withdrawn
LIBRARY

PREFACE

Fuzzy sets were first advocated by Professor L.A. Zadeh in 1965. Besides a few specialists, the world didn't pay much attention to fuzzy sets for the first 10 years, but recently there has been a rapid growth in the number of researchers and papers devoted to them. The field has developed so far as to establish the International Fuzzy Systems Association (IFSA).

On the other hand, it is said that three conditions are necessary for the establishment of a new field: first, a societal need; second, a new methodology (both ideas and techniques); and third, attractiveness to researchers. Let's see how well fuzzy set theory fulfills these conditions.

The first condition is need, which is clearly felt if looked at from engineering and human points of view. Advanced technology is intimately connected with people's lives; the connection between artificial intelligence and "thought," which is the essence of humanity, by itself has a great influence. For everyday people to master artificial intelligence, it is necessary that computers understand the language of humans. The problem is that there is a great deal of ambiguity in everyday language, and it cannot be handled by standard logical processing. A new, logical tool that can express ambiguity is necessary, and fuzzy sets can be considered appropriate in this case. Futhermore, the problem of harmony between people and

ix

technology can become extremely complicated, requiring no collaboration between the natural sciences and the humanities and social sciences, fields that have up until now progressed independently. Fuzzy sets are a communication medium that speaks to both the logical nature of the sciences and the complexity of the humanities and social sciences. We can conclude that the societal need for fuzzy sets will grow in the future.

What about the second condition, for a new methodology? Fuzzy set theory, which was first seen only as a technique for mathematically expressing linguistic ambiguity, is now established as a mathematical measure of a wide variety of ambiguous phenomena, including the concept of probability. Furthermore, a logical system for ambiguity is being established; no points of ambiguity remain in fuzzy set theory. If this advance is made use of, an unstrained theoretical development that comprises opposing concepts, such as subjectivity and objectivity, vagueness and precision, macro and micro viewpoints, and emotion and logic, becomes possible, and makes for a unique methodology.

The third condition is for an attractive field of study, and this has both theoretical and practical sides. As a mathematical system, fuzzy sets expand current frameworks and build a world that takes in new concepts, so they have interested researchers on the theoretical side from early on. On the other hand, there was no good example of practical use for a long time, but with utilization in control systems and artificial intelligence, fuzzy sets have begun to interest practical planners.

The three conditions above are somewhat complementary. That is, if a new methodology is developed, a desire to make some kind of use of it will come along. And if examples of practical use grow, the interest of society will be aroused and needs will arise. That process becomes the stimulus for further theoretical development. The recent, sudden expansion of fuzzy sets means that they have come to this point.

Several books explaining fuzzy sets have already been published. However, most of them have centered on theoretical explanations, and it has been difficult for those who want to learn about fuzzy theory for practical purposes to become familiar with it. Those who want to make use of fuzzy sets as a tool are more interested in its characteristics, problems, comparisons with other methodologies, rules of operations, and hardware than in strict theory. In consideration of these needs, this book was planned to be a reference book for those aiming at practical use. In contrast to theory, research into practical uses has no framework. Since the practical applications of fuzzy theory extend to a wide variety of fields, the processes of analysis, interpretation, and establishment of mathematical models are different for each problem. In other words, each one is an independent creation. Therefore, this book takes the form of a collection of examples, and theory is used only to the extent needed to explain the situations. We hope that the readers of this book understand the motivations of the researchers, the circumstances of the problems, and the features of fuzzy set theory through these case studies and are able to apply them in their own work.

Happily, Japan is in the first rank when it comes to practical applications of fuzzy sets, and there is no lack of examples. This book may prove an opportunity to provide a stimulus to the societal needs for fuzzy sets, and thus research will grow wider and deeper.

Toshiro Terano
for the editors

List of Contributors

Numbers in parentheses indicate chapters and chapter sections to which the authors contributed.

Kiyoji Asai, Dr. *Osaka Institute of Technology, (1-3)*

Kaoru Hirota, Dr. *Hosei University (11-2, 12-1, 12-2, 13)*

Mitsuru Ishizuka, Dr. *The University of Tokyo (16)*

Sosuke Iwai, Dr. *Kyoto University (15)*

Shigehiro Masui, Dr. *Hosei University (12-3)*

Masaharu Mizumoto, Dr. *Osaka Electro-Communication University (3)*

Tetsuji Okuda, Dr. *Osaka Institute of Technology (5)*

Michio Sugeno, Dr. *Tokyo Institute of Technology (1-2, 2, 8, 10)*

Hideo Tanaka, Dr. *University of Osaka Prefecture ((4, 7)*

Toshiro Terano, Dr. *Laboratory for International Fuzzy Engineering Research (1-1, 11-1, 11-3)*

Yahachiro Tsukamoto, Dr. *Meijo Junior College (9)*

Motohide Umano, Dr. *Osaka University (14)*

Junzo Watada, Dr. *Osaka Institute of Technology (6)*

Chapter 1

OUTLINE

1.1 THE HOW, WHAT, AND WHY OF CONVERSION TO FUZZY SYSTEMS

Fuzzy sets are a mathematical concept proposed by Prof. L. A. Zadeh in 1965, but in the background one can see a concealed wish to improve the relationship between humanity and the computer. If there are points of similarity between computers, which are logical machines, and the thinking of people, with their emotions and intuition, there are also differences. If the capabilities of humans and computers could be put together, a remarkable system would be possible. Whether or not fuzzy sets can serve as a go-between depends on the extent of their applications.

Putting aside the purely theoretical study of fuzzy sets, it is always necessary to make comparisons with other methodologies when we think about applications. If there are no results that cannot be obtained from another method, even if those similar results are obtained with simpler processing, it is difficult for a new methodology to survive. Above all else, the outstanding feature of fuzzy sets is the ability to express the amount of ambiguity in human thinking and subjectivity (including natural language) in a comparatively undistorted manner. Therefore, the fields of application involve

1

problems connected to the very heart of humankind. If we divide these problems into three types of systems—machine, human, and human-machine, we come up with the following.

Machine Systems

The first problem with machine systems is that of giving machines a high level of intelligence. Requirements have developed for a high level of circumstantial judgment and complex automatic functions in everything from robots, artificial intelligence, and automated machinery to general machinery. At present, intelligence is expressed in terms of propositions in knowledge engineering, where recognition, judgment, evaluation, and inference are being imitated. However, human thinking includes illogical elements like intuition and inspiration, and it is virtually impossible to express these in terms of two-valued logic. They probably work by bringing together the mass of knowledge called common sense and the metaknowledge that retrieves it. If a machine could do this, it would have achieved human knowledge and ways of thinking.

Human Systems

In contrast to machine-based systems, these try to introduce scientific methodology into the complicated, ambigious problems of man and society. To put it another way, human systems are the building of human models. Because recent technological development has come very suddenly, social customs and institutions have not been able to keep up, and here and there distortion has arisen. Information technology, especially, is connected with human psychological activity, and has a great influence. In this case research into the human side of things is important, but up to now modeling has been an application of either extremely abstract written descriptions or pure mathematical models based on unnatural assumptions. If human problems are modeled on ambiguous forms, research in this field should accelerate.

Human/Machine Systems

These systems use both people and machines in collaboration to get work done that cannot be done by either alone. Knowledge engineering, CAD (computer aided design), and office automation are typical examples. In these kinds of systems, the communication between human and machine is extremely important, and if the communication is poor, even the most fabulous machine will not help. The most important problem to be solved is the simplification of the input process from people to machines. Up to now, people have worked by bringing themselves to the level of machines, but in a high-level human/machine system, the level of the machines must rise high enough to make communication easy. In other words, they must be able to understand natural language and pictorial information. Another problem is the communication from the machines to people, and one would hope for information furnished in a way that stimulates man's creativity and synthetic judgment. These types of communication problems are extremely ambiguous.

Concrete examples of the various problems described above are shown in Table 1.1, which reflects the current research and research themes considered desirable by members of the Japan chapter of IFSA (International Fuzzy Systems Association), according to the results of a 1985 survey.

What are the troublesome points if current scientific and technological techniques are used on the above problems? Let us look at this question from a modeling point of view. Many kinds of models are used in science and technology, but the clearest are those mathematical models that follow physical

Table 1.1.   Survey of Utilization Research on Fuzzy Sets

Machine Systems	Human-Based Systems	Human/Machine Systems
picture/voice recognition	human reliability model	medical diagnosis
Chinese character recognition	cognitive psychology	inspection data processing
natural language understanding	thinking/behavior models	transfusion consultation
intelligent robots	sensory investigations	expert systems
crop recognition	public awareness analysis	CAI
process control	risk assessment	CAD
production management	environmental assessment	optimization planning
car/train operation	human relations structures	personnel management
safety/maintenance systems	demand trend models	development planning
breakdown diagnosis	energy analysis	equipment diagnostics
electric power system operation	market selection models	quality evaluation
fuzzy controller	category analysis	insurance systems
home electrical appliance control	social psychology	human interface
automatic operation		management decision-making
		multipurpose decision-making
		knowledge bases
		data bases

laws. Since a strict check as to whether physical laws are followed during a production process is carried out, the interpretation is clear after completion. Furthermore, once completed, it is easy to obtain a quantitative explanation by using the results. However, if people or societies are the objects, no clear laws exist as in physics, and a lot of assumptions must be made to introduce simple laws. A complex problem will necessitate a vast number of assumptions, and with an ambiguous problem it is difficult to decide which assumptions are valid. If the assumptions are too arbitrary, the model gets divorced from reality. Too precise a model is just as bad as one that is too rough.

There are also other mathematical models that do not use physical laws. Data from actual phenomena are collected, and modeling is done through statistical processes. Although this is the method most commonly used, human and societal interpretation of a model generated in this way is not always easy, and there is no guarantee of accuracy. Therefore, even though this type of model is expressed in quantitative terms, it should be understood qualitatively. Also, a large amount of uniform data is needed for statistical models. With a complex, ambiguous object, the word *uniform* has no clear definition, and it is not easy to gather large amounts of data. Even if some data that show the essence of the object are collected, they will be buried by the large amount of worthless data, when statistical operations are performed. But since the essence is still unknown, no data can be removed. In the world of science, decisions about truth cannot be made by majority.

Recently, logical models, which take a different form from mathematical ones, have come into use. One of them is the structural model, wherein the principle factors that make up the problem are determined and the presence or absence of mutual relationships among these factors investigated and represented graphically. This is a qualitative model, but it is effective for use with complex, ambiguous problems because the structural characteristics appeal to our visual sense. Also, even though it is a model, human intuition and insight are necessary for interpretation, as with a written description. However, since it is a graphic representation, there is greater freedom for interpretation, and since the structure itself is objective, it works to bring together subjectivity and objectivity. The problems with structural models are that the definition of the word *relationship* for the principle factors is not clear and that it uses a two-valued, "yes" or "no" system to represent relationships.

Predicate logic offers a different logical model. Used not only in artificial intelligence but also in cognitive psychology, it forms the basis of knowledge engineering. In predicate logic, regardless of subjectivity or objectivity, the knowledge that humans have is expressed by short sentences called *propositions*. The propositions are a *written* model, so they can be used for ambiguous objects. To use these propositions for predicate logic, the definitions and meanings must be strictly determined. The problems here are as would be expected: the strictness and the two-valued logic. This is because the subjects of the propositions are divorced from reality and cannot be called models. Also, there is room for examination of the reasoning and coordination among the propositions themselves.

As a result, one can say that none of the modeling approaches used up to now are very suitable for use with problems with complex ambiguities. At this point we want something that satisfies our requirements for such contradictions as logic and illogic, subjectivity and objectivity, macro and micro, qualitative and quantitative, ambiguity and precision. The way in which humans think incorporates these kinds of contradictions, so it would be ideal if they could be expressed.

Fuzzy set theory meets these requirements to a certain extent. It is especially suitable for expressing the ambiguity of meaning found in natural language, since it loosens up the overly inhuman logicality of predicate logic. However, if the problem taken as an object is not fully understood when it is used, there will be no results. As with other models, there are both good points and shortcomings to fuzzy models. If they are used without considering their shortcomings or when another model would be more appropriate, the result will be that one goes to all the trouble of making the calculations and building the model but is unable to interpret it and will not understand the results.

When we make use of fuzzy sets, we must make the following points clear:

(1) What part of the problem do we fuzzify, and for what purpose?
(2) What kind of fuzzy model will we use?

To understand the features of fuzzy sets as they apply to modeling, it is better to divide our thinking into two categories, "set models" and "fuzziness." Let us first take a look at the problem of modeling with ordinary sets. Since the sets are groups made from the constants, variables, and functions of the object system, the expression is macroscopic and more ambiguous than models that are not based on groupings, just as showing the range in which a numerical value exists is not as strict as just stating a value. If the state and constraints of a system or its input and output and evaluation are expressed as an ordinary set, that in itself introduces one kind of ambiguity. In this case, the "laws" of the system (input/output relationships, constraints, etc.) are the "mappings" that express the relationships between sets. When functions are used for mapping the model comes close to being mathematical, but logical operations are also common, and in this case we get a logic-type model. Other than predicate logic, set models are not very widely known, so their characteristics, meanings, and features must be fully understood. For example, the set elements, the range of their definition, the expression of relationships between sets, what that expression means, etc., must be clear. Even with predicate logic, if the first definitions and understanding are ambiguous and expressed by unsatisfactory sentences, there is a good chance that the results will be misunderstood, because of the flexibility of the sentences.

Next comes the conversion to fuzzy sets. This is done by gradation of the boundaries of the states, relationships, constraints, and goals of the system model made with ordinary sets. Care has to be taken that even though the boundaries are made vague, the logical structure itself is not. Therefore, the effectiveness of the conversion to fuzzy sets must be that the range of definition becomes ambiguous and that gives rise to a haziness in the relationships that go along with them. In this way, the unnaturalness of dividing the meanings of natural language and the truth of propositions into two values is canceled. Making the boundaries of sets vague has a completely different meaning from statistical variation. The former can freely delineate an ambiguous situation using things like individual subjectivity, experiences, and common sense. The ambiguity of language and emotion is of this type. On the other hand, as stated before, the latter expresses what proportion of the total of a large number of data is occupied by a certain type of data. Therefore, when there is little data, when the total number of the data has no meaning, or when the occurrence of events is not clear, probability cannot express the amount of ambiguity. Next we move on to using these fuzzy sets to make a goal of model for which the goals must first be clear. We may choose from among the three fields of application—machine, human and human-machine systems—for which the aims are as follows:

(1) Express human experiences, common sense, etc., in a form that machines can use.
(2) Make models of human feelings or language.
(3) Imitate human pattern recognition, overall judgment, or general understanding.
(4) Convert information into a form that people can easily understand.
(5) Compress large amounts of information.
(6) Make models of human psychology or behavior.
(7) Make models of societal systems.

Once the goal has been decided, the problem arises of what part of the system and what form the conversion to fuzzy sets will take. There are many forms of system models, but most of them include the following: state variables, independent variables, decision variables, disturbance, laws of cause and effect (transition), their truth values, goals, constraints, evaluation functions, various types of constants.

All of the above are ambiguous. However, if all of them are expressed in terms of fuzzy models, the results will not serve as a model. That is because a model gets its meaning from being a concise expression of the essence of a real problem. Also, the various amounts of ambiguity mentioned above are mutually related. For example, if the state variables are ambiguous, there will naturally be ambiguity in an evaluation based on the state, even if the evaluation functions are definitive.

There are two methods for developing fuzzy models: using the laws of cause and effect and using those of transition. The first makes use of the laws for operations with sets, so the rules for composition and reasoning express the relationships among the variables expressed in the sets. The problem of which rules are best used is determined by the ease of operation and the quality of phenomenological expression. The other method uses ordinary equations to express cause-and-effect relationships, and fuzzy sets are used for the variables.* In this case the meaning of the model itself is clear, so the results of conversion to fuzzy sets are easy to explain.

The procedure for developing a fuzzy model ideally follows the two-stage order. Sets and logical relationships are set up first, and afterwards the conversion to fuzzy sets takes place. This approach makes clear the whereabout of problems, the results of conversion to fuzzy sets, and the interpretation of the model.

As we said before, the application of fuzzy sets is firmly tied to human thinking and behavior, and is what could be called *human simulation.* Therefore, the study of humans themselves is very important for getting good results. The greatest goal for the use of fuzzy logic is to take in the good points of human beings and compensate for their shortcomings.

The approach of this chapter is mainly from the modeling point of view, but there are many other problems, such as membership function identification and language conversion, which must be investigated. Even if an effective solution cannot be obtained, these investigations will be helpful for understanding problems, grasping characteristics, recognizing effects, and bringing-out the next generation of problems.

* The extension principle is applied to the operation.

1.2   THE CONCEPT OF FUZZY THEORY

Fuzzy theory is a mathematical theory, and what is called *fuzziness* takes in one aspect of uncertainty. Fuzziness is the ambiguity that can be found in the definition of a concept or the meaning of a word. For example, the uncertainty in expressions like "old person," "high temperature," or "small number" can be called fuzziness.

Up to now probability has been the only uncertainty with which mathematics has worked. The uncertainty of probability generally relates to the occurrence of phenomena, as symbolized by the concept of randomness. For example, "it will rain tomorrow" "roll the dice and get a three' have the uncertainty of phenomenological occurrences.

Randomness and fuzziness differ in nature; that is, they are different aspects of uncertainty. For example, since the uncertainty of "it will rain tomorrow" comes about because of a prediction made before tomorrow arrives, it will be clarified by the passage of time and the arrival of tomorrow. The uncertainty in "roll the dice and get a three" is also a product of guessing before the roll, and if we actually roll the dice and test it, the proposition becomes certain. However, the uncertainty found in "old person" or "high temperature" is not clarified by the passage of time or testing. The ambiguity lies in the meaning of the words, and since it is an essential characteristic of the words, it always follows them around to some extent.

Probability theory is said to have been developed in the 17th century, so it has a long history. From the very beginning, engineering made use of its ideas, and they are widely used in the natural sciences. On the other hand, fuzzy theory was only developed about 25 years ago and its use is not yet widespread; but fuzziness expresses a much more everyday uncertainty than probability. That is because fuzziness expresses the uncertainty that is a part of the meaning of words, and words are indivisible from man's thinking.

All people think and transmit their thoughts and information by means of words. If probability weren't known to the general public through the announcements of the weather bureau, only people who like to gamble and those preparing for entrance examinations would have anything to do with it. However, everyone is involved with fuzziness, and it is a kind of uncertainty that anyone can understand. If this type of uncertainty could be dealt with mathematically and engineering could make use of it, the effects would be immeasurable. It is said that the difference between computers, which can only process two-valued information, and people is that the latter can deal with ambiguity, but now this outstanding human ability can be expressed by fuzzy theory, handed over to computers and applied to engineering.

Aside from conceptual definitions and the meanings of words, some conceive of fuzziness broadly enough to include things like the uncertainty of people's subjective judgments. The well-known *subjectivity factor* expresses judgments of one-time phenomenological occurrences in terms of probability. This wide definition of fuzziness includes probability-type uncertainty as one type of judgmental uncertainty.

The general terms of the fuzzy theory that makes use of fuzziness are *fuzzy set theory*, *fuzzy logic* and *fuzzy measure theory*. Fuzzy set theory expresses fuzziness in the narrow sense by means of the concept of sets. Fuzzy measure theory is a theory that handles fuzziness in the wider sense. Fuzzy logic is the concept of fuzzy sets incorporated into the framework of multivalued logic. There is also what is called "fuzzy mathematics," standard math into which fuzzy sets and the principles of fuzzy measure have been introduced.

1.3　APPLICATIONS NOW AND THE FUTURE OUTLOOK

1.3.1　Overview

Following development of the above concepts, fuzzy set theory was expanded to areas such as measure and logic for systems theory even while the basic theory was being developed,[1] and it was further developed to include methodologies for applications such as modeling, evaluation, optimization, decision making, control, diagnosis, and information.[2-10] Also there has been testing on various real problems like control, artificial intelligence, and management, and fuzzy theory is actually being used in some areas. The applications of fuzzy systems theory are not restricted to its direct introduction, and there are plans for widespread development of the basic concepts of fuzzy theory. Furthermore, the effects of ambiguity are recognized from the standpoint of fuzzy engineering, and this field is actively advancing the incorporation of these concepts.

Since fuzzy systems theory is the starting point for developing models of ambiguous thinking and judgment processes, the following fields of application are conceivable:

(1) the making of human models that can be used for management and societal problems;
(2) imitation of high-level human abilities for use in automation and informational systems;
(3) development of people-oriented interfaces between people and machines;

(4) other societal and artificial-intelligence applications (risk analysis and prediction, development of functional devices).

Following these classifications, Table 1.2 gives an overview of the internal and external fields of application for fuzzy systems theory.

Since the field of control engineering attaches importance to hardware, it is the one field of those in this table that produces the greatest value in terms of salable produces, and it got attention first. There were examples of applications in 1980 both inside and outside Japan, and products are already on the market.[12, 13]

In standard control theory, a mathematical model is assumed for the controlled system, and control laws that minimize the evaluation functions are determined; but when the object is complicated, mathematical models cannot be determined and one cannot figure out how to decide on the evaluation functions. In these cases, skilled individuals perform control functions by using their experience and intuition to judge situations on the basis of what they

Table 1.2. Overview of Applications of Fuzzy Systems Theory

Field Classification	Management/Society	Artificial Intelligence/Information	Control Engineering
(1) human models	planning evaluation decision making organization human relations		
(2) imitating human capability	decision-making support systems medical diagnostic support systems	expert systems data bases	process control training operations casting handling robots
(3) human interface	signs/advertisements equipment for handicapped people	voice input letter/figure pattern recognition display voice output	
(4) other	risk analysis breakdown prediction (nuclear reactors etc.) earthquake prediction	development of reasoning devices	

think best. There are many examples in the operation of special plants and apparatus and in the piloting and operation of aircraft and trains. Fuzzy control is carried out by programming the rules of control used by skilled individuals, $A_i \rightarrow B_i$ ($i = 1, 2, \ldots, n$), in which the judgment of what kind of manipulated valuable B for given state A is best is made by a computer.

In contrast to control, the fields of artificial intelligence and management are not so conspicuous, because they are centered on software, but there have been innumerable studies since fuzzy set theory was first proposed. Artificial intelligence is both an old and a new theme. The old part was characterized by learning and by pattern recognition, but there has been a shift of emphasis to knowledge engineering and expert systems, which reflect the recent AI boom. Fuzzy inference is commonly used in these. In management systems, current techniques such as multivariate analysis, mathematical programming, and statistical decision theory are being expanded by means of fuzzy system theory for fields like merchandise evaluation,[14] optimization, and decision-making. These techniques are being applied to real problems. Fuzzy measure is used here, along with fuzzy set theory.

1.3.2   Organization of Applications Research

The central organizations for applications research are two research committees of the IFSA, Intelligent Systems and Fuzzy Systems in Business and Manufacturing. The goal of the former is intelligent systems like artificial intelligence and robots and that of the latter is development of techniques and control equipment for management and industry. There is another research committee, Fuzzy Mathematics, which works with attempts to develop applications for theoretical results. This is the international organization, but Japan, the United States, Europe and China have their own regional chapters with their own organizations for applications research. To go along with the operations of the above research committees for applications in management and industry, the Japanese branch has five research groups, Fuzzy Reasoning and Expert Systems, Fuzzy Operations Research, Fuzzy Control, applications to civil engineering, and applications to non-engineering fields which work in cooperation with the committees of the international association. Scheduled activities include an annual symposium, eight research meetings a year in the Kanto and Kansai regions, and news publications. The other regional branches carry out their own activities independently, but the Chinese journal *Fuzzy Mathematics*, which was first published in 1981 and comes out quarterly, is especially well known.

Table 1.3. Fields of Fuzzy Systems Theory Applications in Japan (Survey Results) (1984–1985)

	Operations Research	Social Sciences	Fuzzy Control	Artificial Intelligence	Survey Analysis/Planning/ Evaluation	Other
current research and fields of actual use	fuzzy logic fuzzy reasoning fuzzy operations multipurpose planning group decision theory selection theory game theory PERT multivariate analysis/cluster analysis time series analysis	expression of individuality & subjectivity market selection	learning controller (test by N Co.) water purification plant control (F Co. testing) automatic train operation (H Co. testing) optimal route selection for cars and ships operator behavior analysis process controller (F Co. marketing)	form judgment from pictures pattern recognition learning systems waste paper retrieval robot (H Univ. testing) plant diagnosis data bases CAI	assessment and analysis of public awareness about nuclear energy power system development planning overload cancellation water resource problems factory relayout planning safety eval. of nuclear assessemnt power plants environmental assessment	constructing a fuzzy logic circuit (K Univ. testing) (analog) cell automaton

12

future applications	human relations theory social psychology	robot control	natural language comprehension	risk analysis fuzzy data evaluation	human interface
		train schedule control	interpretation of human information processing systems		
		control of "drying" in chemical engineering	intelligent robots		
			robots to care for the aged		
			expert systems		
			breakdown/equipment diagnosis		
			high temp. furnace diagnosis		
			medical, pancreatic cancer diagnosis		
			decision support systems		
			information retrieval		
			home advisors		

1.3.3 Current Japanese Research and the Future Outlook

An outline of current Japanese applications research was given in 1.3.1 and 1.3.2, but more detailed data can be found in Table 1.3, which shows the results of a survey taken by the Japanese branch of the IFSA. It shows the responses to questions as to fields in which applications research and actual applications are being carried out, as well as fields with future possibilities.

The construction of the fuzzy logic circuit in the "other" column of Table 1.3 is very important for the inference operations that go on in control and expert systems. Inference operations will be speeded up and actualization pushed forward by the development of new electronic devices of this nature.

The human interface noted in the "other future applications" area can also be considered an appropriate field of application when thought of in terms of the original human thinking and judgment modeling focus of fuzzy systems.

Table 1.4. Human Interface and Concepts and Objects of Fuzzy Systems Theory

People	Interface	Tools/System	Objects of Fuzzy System Theory or Concepts
operators	program keyboard display	various kinds of VDT	function apportionment programming communication work evaluation human error interpretation (vigilance)
drivers	program operation section instruments	various kinds of transportation	function apportionment ATC programs communication human error analysis (vigilance work)
homemakers	program operating section display	home electronics	improvement of operation methods ease of use and appeal
handicapped	program operation section display	welfare equipment	function apportionment (supplemental) programming communication feeling

Table 1.4. (*Continued*)

People	Interface	Tools/System	Objects of Fuzzy System Theory or Concepts
managers doctors	program keyboard display	expert systems	input/output communication handling of experience information
passengers	program various signals-guided information various displays	transportation systems (entrance, ticket collection, signals, etc.)	human behavior layout, feeling, system maintenance
customers salespeople	program keyboard display	business systems (vending machines, POS, etc.)	human behavior layout, feeling, reliability, advertisement, will to purchase
employees students	educational materials keyboard display	educational systems (CAI, simulators, etc.)	human engineering (learning, training), educational materials, evaluation
general public	program police information signals, displays	disaster prevention systems	human behavior (shelter) layout, safety evaluation, training
members	rules information	organization	function apportionment rules communication feeling

Making the interface come about smoothly between people and the hardware and software for various equipment will be very important in advancing fields like information production and automation. Table 1.4 shows current areas of interest for human interface and potential objects for the concepts of fuzzy systems theory.

1.3.4 Current Overseas Research and Future Outlook

As outlined in 1.3.1, emphasis has been placed on the three fields of control, artificial intelligence, and management; other uses for fuzzy systems include such areas as breakdown prediction for nuclear reactors in Europe and earthquake forecasting in China.

In the field of control systems, a company in Denmark developed what can be considered the pioneer, a fuzzy controller for use in cement kilns, which is currently in operation in the United States and Europe. In artificial intelligence, a company in the United States has marketed a medical diagnostic support system, and several other companies are following suit. There is progress being made on decision-making support systems in both the United States and Europe. In the United States, which is the software leader, applications in information engineering will probably accelerate from now on. In addition, a company in the United States has developed an electronic inference device and has announced that it has an inference speed 10,000 times probably advance the actualization of fuzzy control and artificial intelligence to a new level.

REFERENCES

(1) Asai, K. and Negoita, C. V., eds, *Introduction to Fuzzy Systems Theory*, Ohmsha, Tokyo (1978) (in Japanese).

(2) Nishida, T. and Takeda, E., *Fuzzy Sets and Their Applications*, Morikita Shuppan, Tokyo (19XX) (in Japanese).

(3) Wang, P. P., ed., *Advances in Fuzzy Sets, Possibility Theory and Applications*, Plenum Press, New York (1983).

(4) Kandel, A., *Fuzzy Techniques in Pattern Recognition*, John Wiley, New York (1982).

(5) De Mori, R., *Computer Models of Speech Using Fuzzy Algorithms*, Plenum Press, New York (1983).

(6) Kacpryzyk, J., *Multistage Decision Making under Fuzziness*, Verlag TUV Reinland (1983).

(7) Zimmerman, H. J., Zadeh, L. A., and Gaines, B. R., eds., *Fuzzy Sets and Decision Analysis*, North-Holland (1984).

(8) Mamdani, E. H., and Gaines, B. R., *Fuzzy Reasoning and Its Applications*, Academic Press, Cambridge, Mass. (1981).

(9) Prade, H., and Negoita, C. V., *Fuzzy Logic in Knowledge Engineering*, Verlag TUV Reinland (1983).

(10) Negoita, C. V., *Expert Systems and Fuzzy Systems*, Benjamin/Cummings Publishing Company, Inc. (1985).

(11) Terano, Toshiro, ed., *Introduction to Fuzzy Engineering*, Kodansha, Tokyo (1981) (in Japanese).

(12) Zadeh, L. A., "Practical Application of Fuzzy Systems Theory Has Just Started," *Nikkei Electronics*, **12** (1984) (in Japanese).

(13) Narita, M., and Kamichika, H., "Fuzzy System Researches Are Taking a Step toward Realization of a Fuzzy Computer with Intuition, "*Nikkei Computer*, **9** (1985) (in Japanese).

(14) Asai, K., Acquisition, Processing and Analysis of Data for Preferences for Consumer Products and New Products—A Human Engineering Approach," *Studies in Product Development*, **14**, pp. 13–23 (1982) (in Japanese).

Chapter 2

THE BASICS OF FUZZY THEORY

2.1 QUANTIFICATION OF AMBIGUITY

Most natural language contains ambiguity and multiplicity of meaning. The objects of adjectives, especially, are not clearly specific, and are ambiguous in terms of breadth of meaning. For example, if we say "tall person," we cannot clearly determine who is tall or who is not tall. The ambiguity of "old person" comes from the adjective "old." Words are usually qualitative, but ones like "tall" and "old" are perceived in connection with amounts of height or age. If we leave out abstract adjectives like "ambiguous," "vague," and "uncertain," adjectives that involve amounts are common. In engineering, especially, the adjectives that describe the states and conditions of various things are almost always connected to amounts in this way.

Let's take a look at the ambiguity of the meanings of "tall" and "old" in terms of the expression of amount. With a range of height of 140 cm to 200 cm, the degree to which height x [cm] can be called "tall" is μ, that is, we make height x correspond to degree μ ($0 \leq \mu \leq 1$). If the horizontal axis is x and the vertical axis is μ, the graph would be drawn as in Fig. 2.1. This graph expresses the ambiguity of "tall" in terms of quantity. In the same way the ambiguity of the concept of "old" is expressed in Fig. 2.2. The amount on the

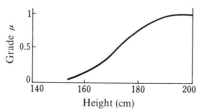

Fig. 2.1. Grade of "Tall"

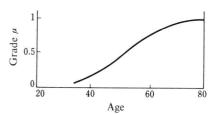

Fig. 2.2. Grade of "Old"

horizontal axis is of course age in years, and the vertical axis shows the degree to which a person can be labeled "old."

As can be seen from these two graphs, the horizontal axis is the quantification of the word, the expression of height or age in one-dimensional space, and the vertical axis is the quantification of the degree of ambiguity. This kind of representation of a word is called *quantification of meaning*. The graph is sometimes called the *quantified meaning*. The meaning of the word is quantified over a specific range; for height this range is 140 cm to 200 cm, and for age it is 20 to 80 years.

Most adjectives are quantified by means of a dimension of meaning like height, age, or width, but abstract numbers such as "a small number" or "a large number" can also be dimensionally quantified. In addition, the concept of "intimate friends" within a group of ten people can be quantified. In this case the names of the 10 people are placed along the horizontal axis; there is no connection with amount.

Fuzzy set theory works with quantification of the meanings of words in graphs within the framework of set theory. It is an attempt to express "tall" or "old" by means of the concept of sets. From a practical point of view, the concept of sets is not necessary for the quantification of ambiguity, but it is possible to increase the range of utilization by working within the framework of set theory. This is because set theory is a very basic concept and has connections with all fields of contemporary mathematics.

Here are some of the essential notations for the introduction of fuzzy sets:

X whole set

E subset of X

$\varnothing$ empty set

$\{0, 1\}$ the set of zero and 1

$[0, 1]$ the real-number interval from zero to 1

χ_E characteristic function of set E

$a \wedge b$ the min of a and b

$a \vee b$ the max of a and b

2.2 FUZZY SETS

An abstract representation of the fuzzy subset of set X would come out something like Fig. 2.3. The rectangular frame represents set X, the dotted circle the ambiguous border of what is inside and outside of it and is $\underset{\sim}{A}$, a fuzzy subset of X. Fuzzy set theory defines the degree to which element x of set X is included in this subset. The function that gives the degree to which it is included is called the *membership function*. The degree of inclusion is sometimes called the *extent* or *grade*. The member is the element x. For example, the grade of membership of element x in area $\underset{\sim}{A}$ is expressed by

$$\mu_{\underline{A}}(x_1) = 1, \qquad \mu_{\underline{A}}(x_2) = 0.8,$$

$$\mu_{\underline{A}}(x_3) = 0.3, \qquad \mu_{\underline{A}}(x_4) = 0$$

etc. μ is the membership function and gives the membership grade, a value from zero to 1. The subscript of μ, $\underline{A}$, shows that $\mu_{\underline{A}}$ is the membership function of $\underline{A}$.

What has gone on here, assuming an ambiguous area $\underline{A}$ as in Fig. 2.3, is an attempt to define quantitatively the ambiguity of $\underline{A}$ by means of the degree $\mu_{\underline{A}}(x)$, to which the element x of set X is included in $\underline{A}$. Mathematically, we cannot assume the existence of an ambiguous area, so this process does not define fuzzy subsets. A formal definition goes as follows.

Fuzzy Set
The function $\mu: X \to [0, 1]$ is given the label $\underline{A}$, and $\underline{A}$ is called a fuzzy (sub)set of X. μ is called the membership function of $\underline{A}$.

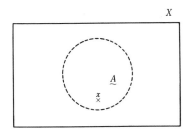

Fig. 2.3. Fuzzy Subset $\underline{A}$

Since a fuzzy set is always defined as a subset of a general set X, the "sub" is frequently abbreviated, and it is just called a fuzzy set. From the definition we see that the function over the interval [0, 1] has a one-to-one correspondence with the fuzzy set. This function is a quantification of the ambiguity of area A; therefore, we can see that it in fact has the same characteristics as the graphs in Figs. 2.1 and 2.2. For example, Fig. 2.1 can be thought of as a representation of the membership function of the "group of heights that can be thought of as tall" fuzzy set within the set of heights of 140 cm to 200 cm. In the same way, the graph in Fig. 2.2 can be thought of as the membership function of the fuzzy set which is "the group of ages that can be considered old" in the range from 20 to 80 years. It is a lot of trouble to use these long labels, so we express the ideas simply with terms like "tall" and "old." However, as is clear from the definition, there are an infinite number of fuzzy sets; any form of membership function is possible, so fuzzy sets do not always have to correspond to words.

If we think about the membership functions of the fuzzy sets not only for "tall," but also for "about average" and "short," we come up with something like Fig. 2.4. As can be seen from the figure, there are two basic things that control the fuzzy sets. The first is the horizontal axis, that is, the whole set X. X is called the support set of the fuzzy set, or simply the support. The second is the membership function, For example, anyone would probably think of the membership function of "about average" as rising in the middle, but the grade of about 150 cm or 170 cm would probably vary subjectively with the person doing the thinking. In this way, fuzzy sets can be seen as being subjective, as opposed to standard sets, which are objective.

As has already been shown, fuzzy sets are indicated by a $\sim$ mark underneath the general symbol for sets. This is because fuzzy sets are extensions of standard sets. In fuzzy theory, standard sets are viewed as exceptional cases of fuzzy sets; therefore, the A for the fuzzy set is sometimes replaced by a simple A. When it is especially important to distinguish between fuzzy and standard sets, standard sets are called *crisp sets* or *nonfuzzy sets*. The word

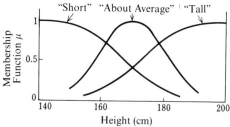

Fig. 2.4. "Short," "About Average," "Tall"

crisp indicates clearly defined boundaries. Let's take a look at the difference between crisp and fuzzy sets in terms of membership functions.

If E is a crisp subset of X, the function

$$\chi_E(x) = \begin{cases} 1; & x \in E \\ 0; & x \notin E \end{cases}$$

is called the *characteristic function* of E. This corresponds to the membership function of E. The grade is two-valued; if x is included in E it is 1; if not zero. In other words, the characteristic function of the crisp set is expressed by

$$\chi_E: X \to \{0, 1\},$$

and the range of values $\{0, 1\}$ is one part of the range of values of the membership function of the fuzzy set. In other words, the heights from 140 cm to 200 cm are divided into three parts, 140 up to but not including 160, 160 up to but not including 180, and 180 and above. The membership functions of these subsets are like those shown in Fig. 2.5. A comparison with Fig. 2.4 makes the differences clear.

The grade in a fuzzy set can be anything from zero to 1, and this range is what makes it different from a crisp set. Each of the crisp subsets of X can be shown to have a one-to-one correspondence with the characteristic function, and because the membership functions are extensions of the characteristic functions, fuzzy sets are extensions of crisp sets. However, we do not have enough facts here to say that they are mathematical extensions. We must also think about operations with sets.

2.3   CRISP SETS

There are three basic operations for crisp sets: operations that give unions, operations that give intersections, and operations that give complements. For

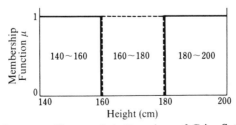

Fig. 2.5.   Characteristic Functions of Crisp Sets

example, the union of sets E and F is defined by

$$E \cup F = \{x \mid x \in E \text{ or } x \in F\}.$$

However, since fuzzy sets are defined by membership functions, the union of fuzzy sets cannot be defined this way. We must use the membership function.

Union

The union of fuzzy sets $\underset{\sim}{A}$ and $\underset{\sim}{B}$, $\underset{\sim}{A} \cup \underset{\sim}{B}$, is the fuzzy set defined by the following membership function:

$$\mu_{\underset{\sim}{A}\cup\underset{\sim}{B}}(x) = \mu_{\underset{\sim}{A}}(x) \vee \mu_{\underset{\sim}{B}}(x). \tag{2.1}$$

Intersection

The intersection of fuzzy sets $\underset{\sim}{A}$ and $\underset{\sim}{B}$, $\underset{\sim}{A} \cap \underset{\sim}{B}$, is the fuzzy set defined by the following membership function:

$$\mu_{\underset{\sim}{A}\cap\underset{\sim}{B}}(x) = \mu_{\underset{\sim}{A}}(x) \wedge \mu_{\underset{\sim}{B}}(x). \tag{2.2}$$

Complement

The complement of fuzzy set $\underset{\sim}{A}$, $\overline{\underset{\sim}{A}}$, is the fuzzy set defined by the following membership function:

$$\mu_{\overline{\underset{\sim}{A}}}(x) = 1 - \mu_{\underset{\sim}{A}}(x). \tag{2.3}$$

Figs. 2.6 and 2.7 are graphs of these. These definitions are also extensions of crisp sets. If χ_E and χ_F are the characteristic functions of crisp sets E and F, and we define χ_1, χ_2 and χ_3 as

$$\chi_1(x) = \chi_E(x) \vee \chi_F(x)$$

$$\chi_2(x) = \chi_E(x) \wedge \chi_F(x)$$

$$\chi_3(x) = 1 - \chi_E(x),$$

we get plots like those in Figs. 2.8 and 2.9 in which χ_1 is the characteristic function for $E \cup F$, χ_1 for $E \cap F$ and χ_3 for $\overline{E}$. From these we can see that the

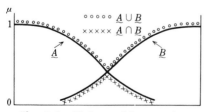

Fig. 2.6. Union and Intersection

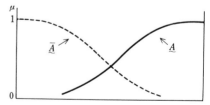

Fig. 2.7. Complement

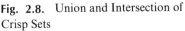

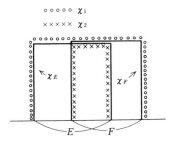

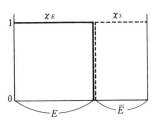

Fig. 2.8.   Union and Intersection of Crisp Sets

Fig. 2.9.   Complement of a Crisp Set

three operations defined for fuzzy sets are extensions of the operations for crisp sets.

Let us discuss two relations of fuzzy sets here.

Equivalence Relation

$$\underset{\sim}{A} = \underset{\sim}{B} \leftrightarrow \mu_A(x) = \mu_B(x); \qquad {}^\forall x \in X. \tag{2.4}$$

Inclusion Relation

$$\underset{\sim}{A} \subset \underset{\sim}{B} \leftrightarrow \mu_A(x) \leq \mu_B(x); \qquad {}^\forall x \in X. \tag{2.5}$$

In general, when we say *extension*, we mean that in most cases some of the original characteristics are lost; thus, fuzzy complements do not always have all of the characteristics of crisp complements.

The complements of fuzzy sets give rise to the following characteristics.

Double Negation Law

$$\overline{\overline{\underset{\sim}{A}}} = \underset{\sim}{A}. \tag{2.6}$$

De Morgan's Laws

$$\overline{\underset{\sim}{A} \cup \underset{\sim}{B}} = \overline{\underset{\sim}{A}} \cap \overline{\underset{\sim}{B}} \tag{2.7}$$

$$\overline{\underset{\sim}{A} \cap \underset{\sim}{B}} = \overline{\underset{\sim}{A}} \cup \overline{\underset{\sim}{B}}. \tag{2.8}$$

But they do not give rise to the excluded-middle law or to the law of contradiction of crisp sets. In other words,

$$\underset{\sim}{A} \cup \overline{\underset{\sim}{A}} \neq X \tag{2.9}$$

$$\underset{\sim}{A} \cap \overline{\underset{\sim}{A}} \neq \varnothing. \tag{2.10}$$

are true. In this connection, the membership function of the empty set $\varnothing$ for all values of x is defined by

$$\mu_X(x) = 1 \tag{2.11}$$

$$\mu_\phi(x) = 0. \tag{2.12}$$

Crisp sets give rise to

$$E \cup \bar{E} = X \text{ (excluded-middle law)}$$

and

$$E \cap \bar{E} = \phi \text{ (law of contradiction)}.$$

The excluded-middle law is a necessary characteristic for two-valued logic. It means that bringing together E and not-E gives the whole; that is, this law maintains that there is no third thing that exists between E and not-E. Since the law of contradiction is derived from the excluded-middle law using the double negation law and De Morgan's laws, it is similar to them. In fact, E is the complement of E because of these characteristics. Therefore, the complement of fuzzy set $\underset{\sim}{A}$ is not a complement in the strict sense. Based on the above, when there is no special need to distinguish them from crisp sets, fuzzy sets will be labeled simply A, B, etc.

2.4   OPERATIONS WITH FUZZY SETS

Many other operations for fuzzy sets have been developed, but we will talk only about the main ones at present.

Algebraic Sum

$$A \boxplus B \leftrightarrow \mu_{A \boxplus B}(x) = \mu_A(x) + \mu_B(x) - \mu_A(x) \cdot \mu_B(x). \tag{2.13}$$

Algebraic Product

$$A \cdot B \leftrightarrow \mu_{A \cdot B}(x) = \mu_A(x) \cdot \mu_B(x). \tag{2.14}$$

Bounded Sum

$$A \oplus B \leftrightarrow \mu_{A \oplus B}(x) = (\mu_A(x) + \mu_B(x)) \wedge 1. \tag{2.15}$$

Bounded Difference

$$A \ominus B \leftrightarrow \mu_{A \ominus B}(x) = (\mu_A(x) - \mu_B(x)) \vee 0. \tag{2.16}$$

λ-Complement

$$\bar{A}^\lambda \leftrightarrow \mu_{\bar{A}^\lambda}(x) = \frac{1 - \mu_A(x)}{1 + \lambda\mu_A(x)}; \qquad -1 < \lambda < \infty. \qquad (2.17)$$

De Morgan's laws apply to the algebraic sum and difference and the complement. In other words,

$$\overline{A \boxplus B} = \bar{A} \cdot \bar{B} \qquad (2.18)$$

$$\overline{A \cdot B} = \bar{A} \boxplus \bar{B} \qquad (2.19)$$

are true. Also, in the case of λ-complements, De Morgan's laws

$$(\overline{A \cup 4})^\lambda = \bar{A}^\lambda \cap \bar{B}^\lambda \qquad (2.20)$$

$$(\overline{A \cap B})^\lambda = \bar{A}^\lambda \cup \bar{B}^\lambda \qquad (2.21)$$

and the double negation law

$$(\overline{\bar{A}^\lambda})^\lambda = A \qquad (2.22)$$

apply. λ is a parameter that gives the *degree of complementation*. If $\lambda = 0$, A^λ is the ordinary complement, and as $\bar{\lambda}$ approaches -1, it approaches the whole set X. When λ approaches infinity, $\bar{A}^\lambda$ approaches the empty set. As can be understood from the characteristic equation, if $\lambda = -1$, the double negation law no longer applies.

The membership function is used to give expression to a fuzzy set, but when the support set is a finite set, the following convenient expression exists. We make

$$X = \{x_1, x_2, \ldots, x_n\}.$$

The fuzzy subset A of X is then expressed by

$$A = \sum_{i=1}^n \mu_A(x_i)/x_i \qquad (2.23)$$

$$\triangleq \mu_A(x_1)/x_1 + \mu_A(x_2)/x_2 + \cdots + \mu_A(x_n)/x_n.$$

The elements of the support set are written on the right side of the slash, the grade on the left. The terms for which the grade is zero are eliminated. The + symbol means *or*, so for example, something like

$$a/x_1 + b/x_1 = a \vee b/x_1$$

gives an operation in which the grade is a max when the elements are the same. Thus, the union of two sets is found by connecting them with a $+$. For example, the fuzzy set of what could be called "large" numbers in the range of integers from 1 to 10 is written

$$\text{large} = 0.2/5 + 0.4/6 + 0.7/7 + 0.9/8 + 1/10.$$

"About middle" is expressed by

$$\text{about middle} = 0.4/3 + 0.8/4 + 1/5 + 0.8/6 + 0.4/7.$$

Their union is found in the following way:

$$
\begin{aligned}
\text{large} \cup \text{about middle} &= (0.2/5 + 0.4/6 + 0.7/7 + 0.9/8 + 1/9 + 1/10) \\
&\quad + (0.4/3 + 0.8/4 + 1/5 + 0.8/6 + 0.4/7) \\
&= 0.4/3 + 0.8/4 + 0.2 \vee 1/5 + 0.4 \vee 0.8/6 \\
&\quad + 0.7 \vee 0.4/7 + 0.9/8 + 1/9 + 1/10 \\
&= 0.4/3 + 0.8/4 + 1/5 + 0.8/6 + 0.7/7 \\
&\quad + 0.9/8 + 1/9 + 1/10.
\end{aligned}
$$

If the support set is infinite, the expression can be extended to give

$$A = \int_X \mu_A(x)/x. \tag{2.24}$$

Using this method of expression, the union, intersection and complement can be written directly.

$$A \cup B = \int_X (\mu_A(x) \vee \mu_B(x))/x \tag{2.25}$$

$$A \cap B = \int_X (\mu_A(x) \wedge \mu_B(x))/x \tag{2.26}$$

$$\bar{A} = \int_X (1 - \mu_A(x))/x. \tag{2.27}$$

Following are a few definitions that we will use later.

Normal Fuzzy Set

A fuzzy set with a membership function that has a grade of 1 is called *normal*. In other words,

$$A \text{ is called "normal"} \leftrightarrow \max_{x \in X} \mu_A(x) = 1. \tag{2.28}$$

Convex Fuzzy Set

When the support set is a real number set and the following applies for all $x \in [a, b]$ over any interval $[a, b]$:

$$\mu_A(x) \geq \mu_A(a) \wedge \mu_A(b) \tag{2.29}$$

A is said to be convex. All of the fuzzy sets in Fig. 2.4 are convex.

Direct Product of Fuzzy Sets

When $A \subset X$ and $B \subset Y$, the fuzzy subset $A \times B$ of $X \times Y$ that can be arrived at in the following way is called the *direct product* of A and B.

$$A \times B \leftrightarrow \mu_{A \times B}(x, y) = \mu_A(x) \wedge \mu_B(y). \tag{2.30}$$

2.5 α-CUTS AND THE EXTENSION PRINCIPLE

Three important parts of fuzzy set theory are what are called the resolution principle, the extension principle, and the representation theorem. But in order to discuss these, we must first introduce the concept of α-cuts.

α-Cut

For a fuzzy set A,

$$A_\alpha \triangleq \{x \mid \mu_A(x) > \alpha\}; \qquad \alpha \in [0, 1) \tag{2.31}$$

$$A_{\bar{\alpha}} \triangleq \{x \mid \mu_A(x) \geq \alpha\}; \qquad \alpha \in (0, 1] \tag{2.32}$$

are called the weak α-cut and strong α-cut, respectively.

The term *α-cut* is a general term that includes both strong and weak types. The weak α-cut is also called the *α* level-set. The difference between strong and weak is the presence or absence of the equal sign. If the membership function is continuous, the distinction between strong and weak is not necessary due to the logical development inherent in the α-cut. Calculations with weak α-cuts are easier to deal with. If the support set is a real number set and the membership function is continuous, the weak α-cut of a convex fuzzy set is a closed interval like the one in Fig. 2.10. Making use of α-cuts, the following relational equation is called the *resolution principle*.

Resolution Principle

$$\mu_A(x) = \sup_{\alpha \in [0,1)} [\alpha \wedge \chi_{A_\alpha}(x)]$$

$$= \sup_{\alpha \in (0,1]} [\alpha \wedge \chi_{A_{\bar{\alpha}}}(x)]. \tag{2.33}$$

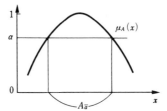

Fig. 2.10. Weak α-Cut

Using the resolution principle, the membership function can be expressed by means of an α-cut. Let us simply describe the proof of expressibility in terms of a weak α-cut. Since $\chi A_{\bar{\alpha}}\alpha(x)$ is the characteristic function of $A_{\bar{\alpha}}$, $x \in A_{\bar{\alpha}}$, that is, $\mu_A(\chi) > \alpha$, and $X \notin A_{\bar{\alpha}}$, that is, $\mu_A(\chi) < \alpha$, it is zero.

Therefore, we get

$$\sup_{\alpha \in (0,1]} [\alpha \wedge \chi_{A_{\bar{\alpha}}}(x)] = \sup_{\alpha \in (0,\, \mu_{A_{\bar{\alpha}}}(x)]} [\alpha \wedge \chi_{A_{\bar{\alpha}}}(x)]$$

$$\vee \sup_{\alpha \in (\mu_A(x),\, 1]} [\alpha \wedge \chi_{A_{\bar{\alpha}}}(x)]$$

$$= \sup_{\alpha \in (0,\, \mu_A(x)]} [\alpha \wedge 1] \vee \sup_{\alpha \in (\mu_A(x),\, 1]} [\alpha \wedge 0]$$

$$= \sup_{\alpha \in (0,\, \mu_A(x)]} \alpha$$

$$= \mu_A(x).$$

If we define the fuzzy set $\alpha A_{\bar{\alpha}}$ here as

$$\alpha A_{\bar{\alpha}} \leftrightarrow \mu_{\alpha A_{\bar{\alpha}}}(x) = \alpha \wedge \chi_{A_{\bar{\alpha}}}(x)], \tag{2.34}$$

the resolution principle is expressed in the form

$$A = \bigcup_{\alpha \in (0,1]} \alpha A_{\bar{\alpha}}. \tag{2.35}$$

In other words, a fuzzy set A is decomposed into $\alpha A_{\bar{\alpha}}$, $\alpha \in (0, 1]$ and is expressed as the union of these. This is what the resolution principle means. It comes out as shown in Fig. 2.11. As can be seen from the figure, if $\alpha_1 < \alpha_2$, $A\alpha_1 \supset A\alpha_2$. Given fuzzy sets such as $\alpha A\alpha_1$ and $\alpha A\alpha_2$, we can retrieve the original membership function of fuzzy set A by connecting the corners of their membership functions. This gives rise to the function

$$\mu_A(x) \leftrightarrow A_{\bar{\alpha}}; \qquad \alpha \in (0, 1].$$

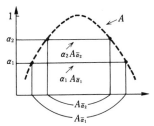

Fig. 2.11. Decomposition of a Fuzzy Set

In other words, a fuzzy set can be expressed in terms of the concept of α-cuts without resorting to the membership function. This is what makes up the representation theorem, and we will leave it at that. α-cuts are very convenient for the calculation of the operations and relations equations equations of fuzzy sets.

Next let us discuss what is called the *extension principle*. We will use the function from X to Y

$$f: X \to Y.$$

This means that the element x of X corresponds to the element $f(x)$ of Y, and if x is a variable within the crisp subset E of X, the range over which $F(x)$ varies is given by

$$f(E) \triangleq \{y \mid y = f(x); \qquad x \in E\}$$

and is called the image of E. The element x inside f is replaced by X, and if we think about this kind of extension of f, we can see that f is the function that makes the subset E of X correspond to the subset $f(E)$ of Y.

Extension Principle
Extending the function $f: X \to Y$, the fuzzy subset A of X is made to correspond to the fuzzy subset $f(A)$ of Y in the following way:

$$\mu_{f(A)}(y) \triangleq \sup_{y=f(x)} \mu_A(x). \tag{2.36}$$

Given y, the right side means that we get the maximum value for $\mu_A(x)$ for x such that $y = f(x)$ (generally there are several). In other words, this is the image of fuzzy set A through f. $f(A)$ can be defined directly as follows:

$$f(A) = \int_Y \mu_A(x)/f(x). \tag{2.37}$$

Because of the characteristics of this method of expression of fuzzy sets, for y, when $x_1 \neq x_2$ and $y = f(x_1) = f(x_2)$, we get

$$\mu_A(x_1)/f(x_1) + \mu_A(x_2)/f(x_2) = (\mu_A(x_1) \vee \mu_A(x_2))/y;$$

therefore, the grade of $f(A)$ for y is exactly like the definition. If f is a one-to-one function x such that $y = f(x)$ is expressed by the inverse function of f, f^{-1}, we get

$$\mu_{f(A)}(y) = \mu_A(f^{-1}(y)).$$

If $f(x)$ varies within the crisp subset F of Y, what is known as the concept of *inverse image* gives the range over which x varies. This is defined by

$$f^{-1}(F) \triangleq \{x \mid f(x) \in F\}.$$

In contrast to the fact that $f(E)$ makes function f act on E, $f^{-1}(F)$ does not make the inverse function f^{-1} act on F. In other words, even if the inverse function does not exist, the inverse image can be defined. The inverse image $f^{-1}(B)$ for fuzzy set B is defined by the following membership function:

$$\mu_{f^{-1}(B)}(x) = \mu_B(f(x)). \tag{2.38}$$

If the α-cut concept is used, we get the following simple expressions for $f(A)$ and $f^{-1}(A)$:

$$f(A)_\alpha = f(A_\alpha) \tag{2.39}$$

$$f^{-1}(B)_\alpha = f^{-1}(B_\alpha). \tag{2.40}$$

The meaning of these equations is clear. For example, to find $f(A)$ for a given A, the strong α-cut $f(A)_\alpha$ is the image $f(A_\alpha)$, through f of strong α-cut A_α (a crisp set) of A. The strong α-cut is used because

$$f(A)_{\bar{\alpha}} \neq f(A_{\bar{\alpha}})$$

is generally true for the weak α-cut. However, if f is a continuous function, it gives rise to an equality for the weak α-cut. Thus, for most cases, the weak α-cut can be used. On the other hand, either the weak or strong α-cut can be used for $f^{-1}(B)$. As has already been shown, the membership function of a fuzzy set can be retrieved given the α-cut, and the use of the α-cut expression for images is convenient for calculations.

The image $f(A)$ and the inverse image $f^{-1}(A)$ of a fuzzy set have exactly the same characteristics as the image and inverse image of the crisp set. Rather than call $f(A)$ the image of A, it is better to view it as an extension of f. For real number sets X and Y, we can consider $f(x) = 2x + 1$. When $x = 3$, the value of f is 7. Now, if x is not just a single value, but ranges, say, from 2 to 4, the value of $f(x)$ ranges from 5 to 9. If we express this in terms of the concept of an image of set E, we replace x with the interval $[2, 4]$ and get

$$f([2,4]) = 2 \times [2,4] + 1$$
$$= [5,9].$$

We get an equation in which an interval replaces the numerical value for x on the right, but this just means that when x varies over an interval, the value of the whole equation varies over an interval, so the calculations are simple. For example, we can think of the equation in these terms:

$$2 \times [2,4] + 1 = [2 \times 2, 2 \times 4] + 1$$
$$= [2 \times 2 + 1, 2 \times 4 + 1]$$
$$= [5,9].$$

What happens to the value of $f(x)$ if we have an ambiguous value such as "about 3" for x instead of the interval $[2,4]$? Numbers like "about 3" and "about 5" are called fuzzy numbers and therefore can be expressed on the real number axis as fuzzy sets.

Fuzzy Numbers
For a normal and convex fuzzy set, if a weak α-cut (α level-set) is a closed interval, it is called a fuzzy number.

Fuzzy numbers have membership functions like the ones in Fig. 2.12. If the membership function is continuous, the weak α-cut will be a closed interval,

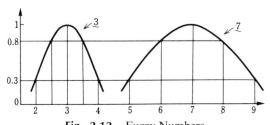

Fig. 2.12. Fuzzy Numbers

but in order to get a closed interval, the membership function does not have to be continuous.

If we express fuzzy numbers like "about 3" and "about 5" as 3 and 5 and replace the x in $f(x)$ with $\underset{\sim}{3}$, the predicted result is

$$f(3) = 2 \times \underset{\sim}{3} + 1$$

$$= \underset{\sim}{7}.$$

If we calculate $f(\underset{\sim}{3})$ using the extension principle, we actually get 7. Let's do the calculation using the α-cut as explained before. Since f is a continuous function, we come up with

$$f(\underset{\sim}{3}_{\bar{\alpha}}) = f(\underset{\sim}{3}_{\bar{\alpha}})$$

$$= 2 \times \underset{\sim}{3}_{\bar{\alpha}} + 1.$$

As shown in Fig. 2.12, 3_α is

$$\underset{\sim}{3}_{\bar{\alpha}} = \begin{cases} 3; & \alpha = 1 \\ [2.5, 3.5]; & \alpha = 0.8 \\ [2, 4]; & \alpha = 0.3. \end{cases}$$

Therefore, we get

$$f(\underset{\sim}{3})_{\bar{\alpha}} = \begin{cases} 7; & \alpha = 1 \\ [6, 8]; & \alpha = 0.8 \\ [5, 9]; & \alpha = 0.3. \end{cases}$$

If we draw the membership function of $f(3)$ using these results, we obtain the fuzzy number we understand to be 7 as shown in the right-hand plot in Fig. 2.12. In this way we can calculate the extensions of functions simply using the concept of α-cuts. In fact, when the support set of the fuzzy set is finite, the calculations are even more simple if done directly using the defining function of the extension principle.

2.6   OPERATIONS WITH FUZZY NUMBERS

Let's take a look at operations with fuzzy numbers as an application of the extension principle. The two-variable function

$$g: X \times Y \to Z,$$

in which $x \in X$ and $y \in Y$ in $g(x, y)$ are replaced by fuzzy sets $A \subset X$ and $B \subset Y$, is defined by

$$\mu_{g(A,B)}(z) = \sup_{z\,=\,g(x,y)} [\mu_A(x) \wedge \mu_B(y)]. \tag{2.41}$$

This expresses the direct product of the A and B in parentheses. The sup is effective in relation to the x and y of $z = g(x, y)$ for a given z. The fuzzy subset $g(A, B)$ of Z is represented by

$$g(A, B) = \int_Z [\mu_A(x) \wedge \mu_B(y)]/g(x, y). \tag{2.42}$$

Using the α-cut, it becomes

$$g(A, B)_\alpha = g(A_\alpha, B_\alpha). \tag{2.43}$$

Finding $g(A, B)$ for

$$g(x, y) = x + y,$$

where x and y are real numbers, corresponds to finding the sum of fuzzy numbers A and B. For example, we might get

$$g(\underset{\sim}{2}, \underset{\sim}{3}) = \underset{\sim}{2} + \underset{\sim}{3}$$

$$= \underset{\sim}{5}.$$

Let's confirm that it actually turns out to be $\underset{\sim}{5}$. For x and y, g is a continuous function, so using the weak α-cut is appropriate:

$$(\underset{\sim}{2} + \underset{\sim}{3})_{\bar\alpha} = \underset{\sim}{2}_{\bar\alpha} + \underset{\sim}{3}_{\bar\alpha}.$$

Since the α-cut of the fuzzy numbers is an interval, the calculation is the sum of the two intervals. Addition of the two intervals $[a, b]$ and $[c, d]$ gives

$$[a, b] + [c, d] = \{\omega \,|\, \omega = u + v; \quad u \in [a, b]; v \in [c, d]\}.$$

In other words, if u and v are numbers that can vary anywhere within the intervals $[a, b]$ and $[c, d]$ respectively, their sum is the range of variation (interval). The result is

$$[a, b] + [c, d] = [a + c, b + d]. \tag{2.44}$$

The α-cut of 3, then, is, as previously noted,

$$\underset{\sim}{3}_{\bar\alpha} = \begin{cases} 3; & \alpha = 1 \\ [2.5, 3.5]; & \alpha = 0.8 \\ [2, 4]; & \alpha = 0.3, \end{cases}$$

and the α-cut of $\underset{\sim}{2}$ is

$$
\underset{\sim}{2}_{\bar{\alpha}} = \begin{cases} 2; & \alpha = 1 \\ [1.5, 2.5]; & \alpha = 0.8 \\ [1, 3]; & \alpha = 0.3, \end{cases}
$$

as shown in Fig. 2.13. So we get

$$
\underset{\sim}{2}_{\bar{\alpha}} + \underset{\sim}{3}_{\bar{\alpha}} = \begin{cases} 5; & \alpha = 1 \\ [4, 6]; & \alpha = 0.8 \\ [3, 7]; & \alpha = 0.3, \end{cases}
$$

and the result, as shown in Fig. 2.13, is the fuzzy number $\underset{\sim}{5}$.

In general, when we think of $g(x, y)$ as a binary operation of x and y like

$$
g(x, y) = x * y,
$$

the expression for fuzzy numbers A and B is

$$
A * B = \int [\mu_A(x) \wedge \mu_B(y)]/(x * y) \tag{2.45}
$$

or

$$
(A * B)_\alpha = A_\alpha * B_\alpha. \tag{2.46}
$$

Here the operation $*$ for both intervals is

$$
[a, b] * [c, d] = \{\omega \mid \omega = u * v; \quad u \in [a, b], \quad v \in [c, d]\}. \tag{2.47}
$$

For example if we look at the calculation of $\underset{\sim}{3} - \underset{\sim}{2}$, where

$$
[a, b] - [c, d] = [a - d, b - c], \tag{2.48}
$$

we get

$$
(\underset{\sim}{3} - \underset{\sim}{2})_{\bar{\alpha}} = \begin{cases} 1; & \alpha = 1 \\ [0, 2]; & \alpha = 0.8 \\ [-1, 3]; & \alpha = 0.3. \end{cases}
$$

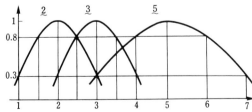

Fig. 2.13. Addition of Fuzzy Numbers $\underset{\sim}{5} = \underset{\sim}{2} + \underset{\sim}{3}$

One note of caution for operations with fuzzy numbers is that subtraction is not the reverse operation of addition. In other words,

$$(3 - 2) + 2 = 3$$

is true, but the case with fuzzy numbers is

$$(\underset{\sim}{3} - \underset{\sim}{2}) + \underset{\sim}{2} \neq \underset{\sim}{3}.$$

If we actually calculate $(3 - 2) + 2$, we get

$$((\underset{\sim}{3} - \underset{\sim}{2}) + \underset{\sim}{2})_{\bar{\alpha}} = \begin{cases} 3; & \alpha = 1 \\ [1.5, 4.5]; & \alpha = 0.8 \\ [0, 6]; & \alpha = 0.3 \end{cases}$$

using our previous results, and the original

$$\underset{\sim}{3}_{\bar{\alpha}} = \begin{cases} 3; & \alpha = 1 \\ [2.5, 3.5]; & \alpha = 0.8 \\ [2, 4]; & \alpha = 0.3 \end{cases}$$

is not completely retrieved. The relationship between multiplication and division is the same.

2.7 FUZZY PROPOSITIONS

Here we will sum up a few of the basic aspects of the fuzzy propositions used in fuzzy logic. Fuzzy propositions are propositions that include fuzzy predications like "it will probably rain tomorrow" and "x is a small number." Generally, they are written

$$x \text{ is } A.$$

A is a fuzzy predicate and is called the *fuzzy variable*. Fuzzy variables are also called linguistic variables and are expressed in terms of fuzzy sets. Without delving into fuzzy logic here, we will discuss the expression of modifications in fuzzy propositions and composite fuzzy propositions.

When the predicate of "x is a small number" is modified to give the form "x is a very small number," we can think in terms of the membership function for the fuzzy set that represents the linguistic variable. This is a proposition modification problem. Words like "extremely" and "very," which change the predicate in this way, are called modifiers, and are indicated by the symbol m. If

$$x \text{ is } A$$

is modified with m, we use the expression

$$x \text{ is } mA.$$

The negation of A, "not," can also be thought of as a modifier. To find the fuzzy sets mA from fuzzy set A we do as follows:

$$\text{very } A = A^2 \tag{2.49}$$

$$\text{more or less } A = A^{1/2} \tag{2.50}$$

$$\text{not } A = 1 - A \tag{2.51}$$

The membership functions of the right-hand terms are

$$\mu_{A^2}(x) = (\mu_A(x))^2 \tag{2.52}$$

$$\mu_{A^{1/2}}(x) = (\mu_A(x))^{1/2} \tag{2.53}$$

$$\mu_{1-A}(x) = 1 - \mu_A(x), \tag{2.54}$$

respectively. These can be shown as something like the plot in Fig. 2.14. The method for representing mA needs to express the meaning of m well, so there is no generalized form.

Next we should discuss composite propositions that are the tying-together of two different propositions, as in "z is a small number, or x is a large number" and "x is small, and is an average-sized number." What connects the propositions is a logical conjunction, and representative examples are "and" and "or." If two propositions are connected we get something like what follows:

$$\ulcorner x \text{ is } A \urcorner \quad \text{or} \quad \ulcorner x \text{ is } B \urcorner = \ulcorner x \text{ is } A \cup B \urcorner \tag{2.55}$$

$$\ulcorner x \text{ is } A \urcorner \quad \text{and} \quad \ulcorner x \text{ is } B \urcorner = \ulcorner x \text{ is } A \cap B \urcorner. \tag{2.56}$$

If we take "x is not very large, it's an average sized number" as an example, we can use the expression "x is $(1 - \text{large}^2) \cap \text{about average}$."

However, composite propositions like "tall" (his height), and "heavy" are

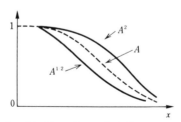

Fig. 2.14. Membership Functions of A, Very A, and More or Less A

not expressed in this way, because the subject is different in each of the two propositions. In this case the expressions are:

$$\ulcorner x \text{ is } C \lrcorner \text{ and } \ulcorner y \text{ is } D \lrcorner = \ulcorner (x, y) \text{ is } C \times D \lrcorner \qquad (2.57)$$

$$\ulcorner x \text{ is } C \lrcorner \text{ or } \ulcorner y \text{ is } D \lrcorner = (x, y) \text{ is } C \times Y \cup X \times D \lrcorner \qquad (2.58)$$

$C \times D$ is the direct product of C and D; X is the support set of C, and Y is the support set of D. This kind of tying-together of two propositions with different subjects (dimensions) is formally expressed by using a fuzzy subset of two dimensions $X \times Y$. Two-dimensional fuzzy subsets are what are called *fuzzy relations* as described in the next paragraph.

A typical combination of fuzzy propositions is "if x is C, y is D." The logic symbol for "if" is an arrow $\rightarrow$, and it is called an *implication*. This proposition can be seen as a two-dimensional composite proposition, and if we write

$$\ulcorner x \text{ is } C \lrcorner \rightarrow \ulcorner y \text{ is } D \lrcorner = \ulcorner (x, y) \text{ is } C \rightarrow D \lrcorner, \qquad (2.59)$$

$C \rightarrow D$ is the fuzzy subset $X \times Y$, and its membership function is expressed by

$$\mu_{C \rightarrow D}(x, y) = (1 - \mu_C(x) + \mu_D(y)) \wedge 1. \qquad (2.60)$$

Aside from "and" and "or" there are many ways to express "$\rightarrow$," and their applications are divided according to the situation.

REFERENCES

(1) Asai, K., and Negoita, C. V., eds., *Introduction to Fuzzy Systems Theory*, Ohmsha, Tokyo (1978) (in Japanese).
(2) Dubois, D., and Prade, H., *Fuzzy Sets and Systems: Theory and Applications*, Academic Press, Cambridge, Mass. (1980).
(3) Sugeno, M., "Fuzzy Theory [I], " *Journal of the Society for Instrument and Control Engineers*, **22** (1), pp. 84–86 (1983) (in Japanese).
(4) Sugeno, M., "Fuzzy Theory [II]," *Journal of the Society for Instrument and Control Engineers*, **22** (4), pp. 42–46 (1983) (in Japanese).
(5) Sugeno, M., "Fuzzy Theory [III]," *Journal of the Society for Instrument and Control Engineers*, **22** (5), pp. 38–42 (1983) (in Japanese).
(6) Sugeno, M., "Fuzzy Theory [IV]," *Journal of the Society for Instrument and Control Engineers*, **22** (6), pp. 50–55 (1983) (in Japanese).
(7) Zimmerman, H. J., *Fuzzy Set Theory—and Its Applications*, Kluwer-Nijhoff Publishing (1985).

Chapter 3

FUZZY RELATIONS

In this chapter we will explain fuzzy relations, which can be discussed as generalizations of ordinary relations. We will begin with a definition of fuzzy relations, talk about expressing fuzzy relations in terms of matrices and graphs, and then define the various operations. Then we will give a somewhat detailed explanation of how fuzzy reasoning, which plays an important role in fuzzy control and fuzzy diagnosis, is explained in terms of compositions of fuzzy relations. In addition, we will talk about fuzzy relational equations and introduce their solutions. Finally, we will discuss similarity relations, similarity classes, and fuzzy order relations, which are special cases of fuzzy relations.

3.1 FUZZY RELATIONS

Ambiguous relationships such as "x and y are almost equal," "x and y look very similar," and "x is much more beautiful than y" are often topics of everyday conversation, but expressing these kinds of ambiguous relationships in terms of ordinary relations is very difficult. *Fuzzy relations* are what makes it possible to express these frequently used ambiguous relationships.

Fuzzy relations can be explained as extensions of ordinary relations, and their range of application is very wide. For example, they are frequently applied in clustering, pattern recognition, inference, systems, and control. They also have applications in the fields known as "soft sciences," such as psychology, medicine, economics and sociology.

3.1.1 Fuzzy Relations

Fuzzy relation R from set X to set Y(or between X and Y) is a fuzzy set in the direct product $X \times Y = \{(x, y) \mid x \in X, y \in Y\}$, and is characterized by a membership function μ_R:

$$\mu_R: X \times Y \rightarrow [0, 1]. \qquad (3.1)$$

Especially when $X = Y$, R is known as a fuzzy relation on X.

Example 3.1.   Let X be a real number set. For $x, y \in X$, the relation "y is *much larger* than x," $x \ll y$, is a fuzzy relation R and can be characterized by the following membership function:

$$\mu_R(x, y) = \begin{cases} 0; & x \geqq y \\ \dfrac{1}{1 + \left(\dfrac{10}{y - x}\right)^2}; & x < y. \end{cases}$$

Example 3.2.   If x and y are people, relationships like "x and y look alike" and "x is much taller than y" are fuzzy relations.

Fuzzy relation R is expressed as follows by using the notation of fuzzy sets:

$$R = \int_{X \times Y} \mu_R(x, y)/(x, y); \qquad x \in X, \qquad y \in Y. \qquad (3.2)$$

Example 3.3.   Let X be a real number set. If $\approx$ means "x and y are *almost* equal," $\approx$ is

$$\approx = \int_{X \times X} e^{-a|x - y|}/(x, y); \qquad a > 0.$$

As a generalization of fuzzy relations, the *n-ary fuzzy relation R* in $X_1 \times X_2 \times \cdots \times Xn$ is given by

$$R = \int_{X_1 \times X_2 \times \cdots \times X_n} \mu_R(x_1, x_2, \ldots, x_n)/(x_1, x_2, \ldots, x_n); \qquad x_i \in X_i \qquad (3.3)$$

and we get the following membership function.

$$\mu_R: X_1 \times X_2 \times \cdots \times X_n \to [0, 1].$$

When $n = 1$, R is an unary fuzzy relation, and this is clearly a fuzzy set in X_1.
When $n = 2$, we have the fuzzy relations of this chapter.

Other ways of expressing fuzzy relations include matrices and graphs.

3.1.2 Fuzzy Matrices and Fuzzy Graphs

Given finite sets $X = \{x_1, x_2, \ldots, x_m\}$, $Y = \{y_1, y_2, \ldots, y_n\}$, a fuzzy relation in $X \times Y$ can be expressed by an $m \times n$ matrix like the one in Fig. 3.1.

This kind of matrix, which expresses a fuzzy relation, is called a *fuzzy matrix*. Since μ_R has values within the interval $[0, 1]$, the elements of the fuzzy matrix also have values within $[0, 1]$.

In order to express fuzzy relation R in a graph, for $\mu_R(x_i, y_j)$, we make x_i, y_j vertices and add the grade $\mu_R(x_i, y_j)$ to the arc from x_i to y_j. This graph is called a *fuzzy graph*.

Example 3.4. When fuzzy relation R on $X = \{a, b, c\}$ is

$$R = 0.2/(a, a) + 1/(a, b) + 0.4/(a, c) + 0.6/(b, b)$$
$$+ 0.3/(b, c) + 1/(c, b) + 0.8(c, c),$$

(3.4)

the fuzzy matrix and fuzzy graph for R are as shown in Fig. 3.2.

$$R = \begin{bmatrix} \mu_R(x_1, y_1) & \mu_R(x_1, y_2) & \cdots & \mu_R(x_1, y_n) \\ \mu_R(x_2, y_1) & \mu_R(x_2, y_2) & \cdots & \mu_R(x_2, y_n) \\ & & \vdots & \\ \mu_R(x_m, y_1) & \mu_R(x_m, y_2) & \cdots & \mu_R(x_m, y_n) \end{bmatrix}$$

Fig. 3.1. Fuzzy Matrix for Fuzzy Relation R

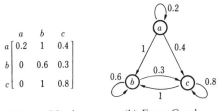

$$\begin{array}{c c c c} & a & b & c \\ a & \begin{bmatrix} 0.2 & 1 & 0.4 \\ b & 0 & 0.6 & 0.3 \\ c & 0 & 1 & 0.8 \end{bmatrix} \end{array}$$

(a) Fuzzy Matrix   (b) Fuzzy Graph

Fig. 3.2. Fuzzy Matrix and Graph for Equation (3.4)

3.2   OPERATIONS FOR FUZZY RELATIONS

Since the fuzzy relation from X to Y is a fuzzy set in $X \times Y$, the operations for fuzzy sets can be used just as is. If R and S are fuzzy relations in $X \times Y$, the operations are as follows:

$$\text{inclusion} \quad R \subseteq S \leftrightarrow \mu_R(x, y) \le \mu_S(x, y)$$

$$\text{union} \quad R \cup S \leftrightarrow \mu_{R \cup S}(x, y) = \mu_R(x, y) \vee \mu_S(x, y)$$

$$\text{intersection} \quad R \cap S \leftrightarrow \mu_{R \cap S}(x, y) = \mu_R(x, y) \wedge \mu_S(x, y)$$

$$\text{complement set} \quad \bar{R} \quad \leftrightarrow \mu_{\bar{R}}(x, y) \quad = 1 - \mu_R(x, y).$$

where $\vee = \max$ and $\wedge = \min$.

Other fuzzy operations can be used for fuzzy relations in the same way.

3.2.1   Composition of Fuzzy Relations

If R is a fuzzy relation in $X \times Y$ and S is a fuzzy relation in $Y \times Z$, the *composition* of R and S, $R \circ S$, is a fuzzy relation in $X \times Z$ as defined below.

$$R \circ S \leftrightarrow \mu_{R \circ S}(x, z) = \bigvee_y \{\mu_R(x, y) \wedge \mu_S(y, z)\}, \tag{3.5}$$

where $\vee = \max$ and $\wedge = \min$. This composition uses max and min operations, so it is called a *max–min composition*.

Example 3.5. Let $X = \{x_1, x_2\}$, $Y = \{y_1, y_2, y_3\}$, $Z = \{z_1, z_2\}$. If the fuzzy relations R and S are expressed by the following matrices, the composition of R and S is found as follows. That is, the composition corresponds to the ordinary product of the matrices, and the product replaces the $\wedge$ (min), the sum the $\vee$ (max):

$$R = \begin{array}{c} \\ x_1 \\ x_2 \end{array} \begin{array}{ccc} y_1 & y_2 & y_3 \\ \left[\begin{array}{ccc} 0.4 & 0.6 & 0 \\ 0.9 & 1 & 0.1 \end{array} \right], \end{array} \quad S = \begin{array}{c} \\ y_1 \\ y_2 \\ y_3 \end{array} \begin{array}{cc} z_1 & z_1 \\ \left[\begin{array}{cc} 0.5 & 0.8 \\ 0.1 & 1 \\ 0 & 0.6 \end{array} \right] \end{array}$$

$$R \circ S = \begin{bmatrix} 0.4 & 0.6 & 0 \\ 0.9 & 1 & 0.1 \end{bmatrix} \circ \begin{bmatrix} 0.5 & 0.8 \\ 0.1 & 1 \\ 0 & 0.6 \end{bmatrix}$$

$$= \begin{bmatrix} (0.4 \wedge 0.5) \vee (0.6 \wedge 0.1) \vee (0 \wedge 0), \\ (0.9 \wedge 0.5) \vee (1 \wedge 0.1) \vee (0.1 \wedge 0), \end{bmatrix}$$

$$\begin{bmatrix} (0.4 \wedge 0.8) \vee (0.6 \wedge 1) \vee (0 \wedge 0.6) \\ (0.9 \wedge 0.8) \vee (1 \wedge 1) \vee (0.1 \wedge 0.6) \end{bmatrix}$$

$$= \begin{array}{c} \\ x_1 \\ x_2 \end{array} \begin{array}{cc} z_1 & z_2 \\ \begin{bmatrix} 0.4 & 0.6 \\ 0.5 & 1 \end{bmatrix} \end{array}.$$

Note: Many other kinds of compositions for fuzzy relations can be considered. For example, there is a *min–max composition* that is the dual composition operation for Equation (3.5), reversing $\vee$ and $\wedge$. This clearly gives rise to

$$R \square S \leftrightarrow \mu_{R \square S}(x, z) = \bigwedge_y \{\mu_R(x, y) \vee \mu_S(y, z)\}$$

$$\overline{R \square S} = \bar{R} \circ \bar{S}.$$

Also, there is the following *max-star composition*:

$$R * S \leftrightarrow \mu_{R * S}(x, z) = \bigvee_y \{\mu_R(x, y) * \mu_S(y, z)\}. \tag{3.6}$$

The $*$ on the right side is defined as a binary operation. If, for example, we use a multiplication dot for $(\cdot)$ for the $*$, we get the *max-product composition*.

3.2.2 Converse Relations

The *converse fuzzy* relation of fuzzy relation R is shown by R^c and defined as

$$R^c \leftrightarrow \mu_R{}^c(y, x) = \mu_R(x, y). \tag{3.7}$$

The following are the most basic fuzzy relations: For any $x, y \in X$,

Identity Relation

$$I \leftrightarrow \mu_I(x, y) = \begin{cases} 1; & x = y \\ 0; & x \neq y. \end{cases}$$

Zero Relation

$$O \leftrightarrow \mu_O(x, y) = 0.$$

Universe Relation

$$E \leftrightarrow \mu_E(x, y) = 1.$$

Example 3.6.   The following are examples of these three relations.

$$I = \begin{bmatrix} 1 & 0 & 0 \\ 0 & 1 & 0 \\ 0 & 0 & 1 \end{bmatrix}, \qquad O = \begin{bmatrix} 0 & 0 & 0 \\ 0 & 0 & 0 \\ 0 & 0 & 0 \end{bmatrix}, \qquad E = \begin{bmatrix} 1 & 1 & 1 \\ 1 & 1 & 1 \\ 1 & 1 & 1 \end{bmatrix}.$$

3.3 BASIC PROPERTIES OF FUZZY RELATIONS

Table 3.1 contains a list of the basic properties of operations with fuzzy relations, especially composition and converse fuzzy relations. Since the properties of unions ($\cup$), intersections ($\cap$) and complements ($^-$) for fuzzy relations are the same as those for fuzzy sets, we will not repeat them here.

3.4 FUZZY RELATIONS AND FUZZY REASONING

We will now try to give a somewhat detailed discussion of fuzzy reasoning (approximate reasoning) as an example of one application of the composition of fuzzy relations "∘" which plays a major role in areas such as fuzzy control, fuzzy diagnosis, and fuzzy expert systems.

Table 3.1.   Basic Properties of Fuzzy Relations

For Composition (∘)	For Inverse Relations (^{c})
(1) $R \circ I = I \circ R = R$	(11) $(R \cup S)^c = R^c \cup S^c$
(2) $R \circ O = O \circ R = O$	$(R \cap S)^c = R^c \cap S^c$
(3) In general, $R \circ S \neq S \circ R$	$(R \circ S)^c = S^c \circ R^c$
(4) $(R \circ S) \circ T = R \circ (S \circ T)$	(12) $(R^c)^c = R, \bar{R}^c = \bar{R}^c$
(5) $R^{m+1} = R^m \circ R, R^c = I$	(13) $R \subseteq S \to R^c \subseteq S^c, \bar{R} \supseteq \bar{S}$
(6) $R^m \circ R^n = R^{m+n}$	
(7) $(R^m)^n = R^{mn}$	
(8) $R \circ (S \cup T) = (R \circ S) \cup (R \circ T)$	
(9) $R \circ (S \cap T) \subseteq (R \circ S) \cap (R \circ T)$	
(10) $S \subseteq T \to R \circ S \subseteq R \circ T$	

First let us take a look at what could be considered a simple example of fuzzy reasoning:

premise 1	If x is A then y is B	
premise 2	x is A'	(3.8)
conclusion	y is B',	

where x and y are objects and A, A' and B, B' are fuzzy sets in the universe sets U, U and V, V, respectively.

In this fuzzy reasoning form, A and A' and B and B' are not necessarily equal, but if $A' = A$ and $B' = B$, the reasoning form in (3.8) is reduced to what is called the *modus ponens*.

The following is an example of a fuzzy reasoning form along the lines of (3.8).

If a tomato is red then the tomato is ripe.

This tomato is very red.

∴ This tomato is very ripe.

Since the fuzzy condition in (3.8), "If x is A then y is B" (for simplicity expressed by $A \rightarrow B$), expresses some kind of relation between A and B, the following is given by Zadeh[2] as a method for translating the fuzzy condition $A \rightarrow B$ into a fuzzy relation following the implication rule $a \rightarrow b = 1 \wedge (1 - a + b)$:

$$R_a = A \rightarrow B$$
$$= (\bar{A} \times V) \oplus (U \times B)$$
$$= \int_{U \times V} 1 \wedge (1 - \mu_A(u) + \mu_B(v))/(u, v).$$

(3.9)

The $\oplus$ means *bounded sum*, which is defined as $a \oplus b = 1 \wedge (a + b)$.

Conclusion B' in (3.8) can be obtained by taking the composition of fuzzy set A' and fuzzy condition $A \rightarrow B$ (the *compositional rule of inference*), and in general is given as follows:

$$B' = A' \circ (A \rightarrow B)$$
$$\mu_{B'}(v) = \bigvee_u \{\mu_{A'}(u) \wedge \mu_{A \rightarrow B}(u, v)\}.$$

(3.10)

For example, with Zadeh's method, R_a is given by

$$B' = A' \circ R_a = A' \circ [(\bar{A} \times V) \oplus (U \times B)]$$

and $$B' = A \circ R_a = \int_V \frac{1 + \mu_B(v)}{2} / v \text{ at } A' = A.$$

However, this case does not satisfy the following *modus ponens*, which is an appropriate requirement for fuzzy reasoning:

$$\begin{array}{c} \text{If } x \text{ is } A \text{ then } y \text{ is } B \\ \underline{x \text{ is } A} \\ y \text{ is } B \end{array} \qquad\qquad (3.11)$$

What, then, would happen with Mamdani's method, which is often used in fuzzy control?

He employs the direct product of A and B for $A \to B$. In other words, we get

$$A \to B = A \times B$$

$$= \int_{U \times V} \mu_A(u) \wedge \mu_B(v)/(u, v).$$

Thus, when $A' = A$, we get the following:

$$B' = A \circ (A \times B)$$

$$\mu_{B'}(v) = \bigvee_u \{\mu_A(u) \wedge (\mu_A(u) \wedge \mu_B(v))\}$$

$$= \bigvee_u \{\mu_A(u) \wedge \mu_B(v)\}$$

$$= \bigvee_u \mu_A(u) \wedge \mu_B(v).$$

If we assume that A is normal (the maximal grade of A is 1), we get the following:

$$= 1 \wedge \mu_B(\vee) = \mu_B(\vee),$$

and we can see that $B' = B$. In other words, we can see that it satisfies the *modus ponens* in (3.11).

Table 3.2 is a list of implication rules frequently used in fuzzy reasoning. R_a and R_m are Zadeh's method, and R_c is Mamdani's method.

Table 3.2. Various Implication Formulae

R_a	$a \to b = 1 \wedge (1 - a + b)$
R_m	$a \to b = (a \wedge b) \vee (1 - a)$
R_c	$a \to b = a \wedge b$
R_s	$a \to b = \begin{cases} 1; & a \leq b \\ 0; & a > b \end{cases}$
R_g	$a \to b = \begin{cases} 1; & a \leq b \\ b; & a > b \end{cases}$
R_b	$a \to b = (1 - a) \vee b$
R_Δ	$a \to b = \begin{cases} 1; & a \leq b \\ \dfrac{b}{a}; & a > b \end{cases}$

Thus far we have used the max–min composition in the compositional rule of inference $B' = A' \circ (A \to B)$, but besides the composition in (3.10) there are others (in general the max–*composition [see Equation (3.6)]) we can consider. Namely, we have

$$B' = A' * (A \to B)$$

$$\mu_{B'}(v) = \bigvee_u \{\mu_{A'}(u) * \mu_{A \to B}(u, v)\}.$$

For example, let the compositions using $\odot$ (bounded-product) and $\wedge$ (drastic product) be used for *, that is,

$$x \odot y = 0 \vee (x + y - 1)$$

$$x \wedge y = \begin{cases} x; & y = 1 \\ y; & x = 1 \\ 0; & x, y < 1, \end{cases}$$

and let the max–$\odot$ and max–$\wedge$ compositions be "$\square$" and "$\blacktriangle$," respectively. If we use these for R_a in Equation (3.9) when $A' = A$, the conclusion B' becomes B:

$$B' = A \,\square\, R_a = A \,\blacktriangle\, R_a = B.$$

In other words, it satisfies the *modus ponens*. Using new compositions like this, we see that we can obtain inference results that match intuition even from the

Table 3.3. Results of Reasoning[4]

Composition $a \to b$ \ A'	Max–Min Composition				Max-Composition				Max-Composition			
	A	Very A	More or Less A	Not A	A	Very A	More or Less A	Not A	A	Very A	More or Less A	Not A
$1 \wedge (1 - a + b)$	$\dfrac{1+\mu_B}{2}$	(*1)	(*3)	1	μ_B	μ_B	(*5)	1	μ_B	μ_B	$\sqrt{\mu_B}$	1
$(a \wedge b) \vee (1 - a)$	$0.5 \vee \mu_B$	(*2)	(*4)	1	μ_B	μ_B	$\dfrac{1}{4} \vee \mu_B$	1	μ_B	μ_B	μ_B	1
$a \wedge b$	μ_B	μ_B	μ_B	$0.5 \wedge \mu_B$	μ_B	μ_B	μ_B	0	μ_B	μ_B	μ_B	0
$\begin{cases}1; & a \le b \\ 0; & a > b\end{cases}$	μ_B	μ_B^2	$\sqrt{\mu_B}$	1	μ_B	μ_B^2	$\sqrt{\mu_B}$	1	μ_B	μ_B^2	$\sqrt{\mu_B}$	1
$\begin{cases}1; & a \le b \\ b; & a > b\end{cases}$	μ_B	μ_B	$\sqrt{\mu_B}$	1	μ_B	μ_B	$\sqrt{\mu_B}$	1	μ_B	μ_B	$\sqrt{\mu_B}$	1
$(1 - a) \vee b$	$0.5 \vee \mu_B$	(*2)	(*4)	1	μ_B	μ_B	$\dfrac{1}{4} \vee \mu_B$	1	μ_B	μ_B	μ_B	1
$\begin{cases}1; & a \le b \\ \dfrac{b}{a}; & a > b\end{cases}$	$\sqrt{\mu_B}$	$\mu_B^{2/3}$	$\mu_B^{1/3}$	1	μ_B	μ_B	$\sqrt{\mu_B}$	1	μ_B	μ_B	$\sqrt{\mu_B}$	1

(note) (*1) $\dfrac{3 + 2\mu_B - \sqrt{5 + 4\mu_B}}{2}$ (*2) $\dfrac{3 - \sqrt{5}}{2} \vee \mu_B$ (*3) $\dfrac{\sqrt{5 + 4\mu_B} - 1}{2}$ (*4) $\dfrac{\sqrt{5} - 1}{2} \vee \mu_B$

(*5) $\begin{cases}\mu_B + \dfrac{1}{4}; & \mu_B \le \dfrac{1}{4} \\[2mm] \sqrt{\mu_B}; & \mu_B \ge \dfrac{1}{4}\end{cases}$

R_a method, which was unsatisfactory when based on the max–min composition. This can also be said for other methods based on the implication rules, as in Table 3.2. This is shown in Table 3.3. In this table, besides $A' = A$, conclusions from

$$A' = \text{very } A = \int_U \mu_A(u)^2/u$$

$$A' = \text{more or less } A = \int_U \sqrt{\mu_A(u)}/u$$

$$A' = \text{not } A = \int_U 1 - \mu_A(u)/u$$

are given.[4] It should be clear that conclusions that match intuition can be obtained based on new composition methods other than $A' = A$ for A'.

The following form of fuzzy reasoning is the one most commonly used in fuzzy control. Here the conditional parts of the fuzzy conditions are tied together with the "and" of two fuzzy propositions:

premise 1	If x is A and y is B then z is C
premise 2	x is A' and y is B'
conclusion	z is C'

Here, $A, A' \subsetneq U$; $B, B' \subsetneq V$; $C, C' \subsetneq W$ are fuzzy sets.

The fuzzy proposition connected by "and," "x is A and y is B," is shown by $A \cap B$ for convenience.

$$x \text{ is } A \quad \text{and} \quad y \text{ is } B \equiv A \cap B,$$

That is, $A \cap B$ expresses the intersection of $A \times V$ and $U \times B$ and equals the direct product $A \times B$.

$$A \cap B = (A \times V) \cap (U \times B)$$

$$= A \times B \tag{3.13}$$

$$= \int_{U \times V} \mu_A(u) \wedge \mu_B(v)/(u, v).$$

The fuzzy condition "If x is A and y is B then z is C" (for simplicity, $A \cap B \to C$) in premise 1 is replaced by the fuzzy relation $R(A, B; C)$ in direct product $U \times V \times W$. For example, noting that for the Mamdani's method R_c in Table 3.2, the extension of the implication $a \to b = a \wedge b^2$ is

$$(a \wedge b) \to c = (a \wedge b) \wedge c,$$

and the fuzzy condition $A \cap B \to C$ is replaced by the ternary fuzzy relation

$$R_c(A, B; C) = \int_{U \times V \times W} \mu_A(u) \wedge \mu_B(v) \wedge \mu_C(w)/(u, v, w). \quad (3.14)$$

In the same way, we get

$$R_a(A, B; C) = \overline{(A \cap B} \times W) \oplus (U \times V \times C)$$

$$= \int_{U \times V \times W} 1 \wedge [1 - (\mu_A(u) \wedge \mu_B(v)) + \mu_C(w)]/(u, v, w)$$

for the R_a method in Table 3.2. Those that follow are found in the same way. The conclusion of equation (3.12) is found in the following way using the compositional rule of inference, where "$\circ$" is the max–min composition.

$$C' = (A' \cap B') \circ R(A, B; C)$$

$$\mu_{C'}(w) = \bigvee_{u,v} \{(\mu_{A'}(u) \wedge \mu_{B'}(v)) \wedge \mu_{R(A,B;C)}(u, v, w)\}.$$

For example, we get

$$C' = (A' \cap B') \circ R_C(A, B; C)$$

$$\mu_{C'}(w) = \bigvee_{u,v} \{(\mu_{A'}(u) \wedge \mu_{B'}(v)) \wedge [\mu_A(u) \wedge \mu_B(v) \wedge \mu_C(w)]\}$$

$$= \bigvee_u \{\mu_{A'}(u) \wedge \mu_A(u) \wedge \mu_C(w)$$

$$\wedge \bigvee_v [\mu_{B'}(v) \wedge \mu_B(v) \wedge \mu_C(w)]\}$$

$$= \bigvee_u \{\mu_{A'}(u) \wedge \mu_A(u) \wedge \mu_C(w) \wedge \mu_{B' \circ R_c(B;C)}(w)\}$$

$$= \mu_{A' \circ R_c(A;C)}(w) \wedge \mu_{B' \circ R_c(B;C)}(w)$$

for R_c. In other words, we get

$$C' = (A' \cap B') \circ R_c(A, B; C)$$

$$= [A' \circ R_c(A; C)] \cap [B' \circ R_c(B; C)].$$

In this instance, $R_c(A; C)$ is the fuzzy relation for condition $A \to C$ obtained by means of method R_c. In other words, with Mamdani's R_c, the conclusion of (3.12) is expressed as the intersection of the individual conclusions. Thus, we have

$$(A' \cap B') \circ (A, B \to C) = [A' \circ (A \to C)] \cap [B' \circ (B \to C)].$$

Example 3.7.   When $A' = A$ and $B' = B$, the conclusion obtained by means of R_c is

$$(A \cap B) \circ (A, B \to C) = [A \circ (A \to C)] \cap [B \circ (B \to C)]$$

$$= C \cap C = C,$$

and the following appropriate reasoning form is satisfied:

$$\text{If } x \text{ is } A \text{ and } y \text{ is } B \text{ then } z \text{ is } C$$

$$\frac{x \text{ is } A \text{ and } y \text{ is } B}{z \text{ is } C} \qquad (3.15)$$

R_a, R_s, R_g, R_b and R_Δ are among the implication rules in Table 3.2 that satisfy

$$(a \wedge b) \to c = (a \to c) \vee (b \to c),$$

but these methods differ from R_c in that the conclusion is expressed by the union ($\cup$) of the individual conclusions:

$$(A' \cap B') \circ (A \cap B \to C) = [A' \circ (A \to C)] \cup [B' \circ (B \to C)].$$

Example 3.8.   When $A' = A$ and $B' = B$, the conclusion of R_a is

$$(A \cap B) \circ (A, B \to C) = \int_W \left. \frac{1 + \mu_C(w)}{2} \right/ w \cup \int_W \left. \frac{1 + \mu_C(w)}{2} \right/ w$$

$$= \int_W \left. \frac{1 + \mu_C(w)}{2} \right/ w (\neq C),$$

and we can see that it does not satisfy Equation (3.15). (Note that we can see that other compositions, such as "$\square$" and "$\blacktriangle$," satisfy (3.15).)

Finally, let us take a look at a slightly more complicated fuzzy reasoning:

premise 1	If x is A_1 and y is B_1 then z is C_1 else
premise 2	if x is A_2 and y is B_2 then z is C_2 else
	$\vdots$
premise n	if x is A_n and y is B_n then z is C_n.

$$(3.16)$$

$$\frac{\text{premise } n+1 \qquad x \text{ is } A' \text{ and } y \text{ is } B'}{\text{conclusion} \qquad\qquad\qquad\qquad\qquad\qquad z \text{ is } C'.}$$

In this case, the conclusion C' using Mamdani's R_c is given by

$$C' = (A' \cap B') \circ [(A_1 \cap B_1 \to C_1) \cup (A_2 \cap B_2 \to C_2)$$

$$\cup \cdots \cup (A_n \cap B_n \to C_n)]$$

$$= [A' \circ (A_1 \to C_1) \cap B' \circ (B_1 \to C_1)]$$

$$\cup \cdots \cup [A' \circ (A_n \to C_n) \cap B' \circ (B_n \to C_n)],$$

where "else" is interpreted as the union ($\cup$).

But "else" is interpreted as the intersection for R_a, R_s, R_g, R_b and R_Δ, which satisfy $(a \wedge b) \to c = (a \to c) \vee (b \to c)$, and conclusion C' is given by

$$C' = (A' \cap B') \circ [(A_1 \cap B_1 \to C_1) \cap (A_2 \cap B_2 \to C_2)$$

$$\cap \cdots \cap (A_n \cap B_n \to C_n)]$$

$$\subseteq [A' \circ (A_1 \to C_1) \cup B' \circ (B_1 \to C_1)]$$

$$\cap \cdots \cap [A' \circ (A_n \to C_n) \cup B' \circ (B_n \to C_n)].$$

These types fuzzy reasoning are frequently used in fuzzy control and fuzzy diagnosis, but, especially in fuzzy control, the A', B' in (3.16) (x is A' and y is B' of premise $n + 1$) is generally a fixed value u_o, v_o rather than a fuzzy set (for example, u_o = error, v_o = change in error).

3.5 FUZZY RELATIONAL EQUATIONS

Let us now discuss fuzzy relational equations, which play an important role in areas such as system analysis, the planning of fuzzy controllers, and the four fundamental fuzzy operations.

As in the previous section, if we let A be a fuzzy set in set X and R be the fuzzy relation in $X \times Y$, the composition of A and R, $A \circ R$, is defined as

$$\mu_{A \circ R}(y) = \bigvee_x \{\mu_A(x) \wedge \mu_R(x, y)\} \tag{3.17}$$

and is a fuzzy set in Y. We call this B. In other words,

$$A \circ R = B \tag{3.18}$$

Example 3.9. If we express fuzzy set A and fuzzy relation R in terms of vectors and matrices and let them be as shown below, the composition of A

and R comes out as follows:

$$A \circ R = [0.3 \quad 0.7 \quad 1] \circ \begin{array}{c} \\ x_1 \\ x_2 \\ x_3 \end{array} \begin{bmatrix} 0.5 & 0.7 & 0.8 \\ 0.6 & 0.6 & 0.4 \\ 1 & 0.5 & 0.3 \end{bmatrix}$$

$$= [1 \quad 0.6 \quad 0.4] = B.$$

If we view A as the fuzzy input, B as the fuzzy output, and R as the fuzzy system, we can think of (3.18) as expressing a fuzzy system with fuzzy input and fuzzy output. In other words, this shows that fuzzy system R gives a fuzzy output of B given a fuzzy input of A. A diagrammatic version would give us something like Fig. 3.3.

If the fuzzy input A and the fuzzy relation R are given, the fuzzy output B can be found simply by taking the composition of A and R. On the other hand, let us take a look at the following problems:

(1) Given A and B, finding R.
(2) Given B and R, finding A.

(1) corresponds to finding the fuzzy system when the fuzzy input and fuzzy output are known, and (2) corresponds to finding the fuzzy input when the fuzzy output and fuzzy system are known.

When we are dealing with these types of problems, (3.18) can be thought of as expressing one type of equation, and from this,

$$A \circ R = B \tag{3.19}$$

is called *fuzzy relational equation*.

In preparation, let us introduce the following operation. For any $a, b \in [0, 1]$, let

$$a \alpha b = \begin{cases} 1; & a \leq b \\ b; & a > b \end{cases} \tag{3.20}$$

Fig. 3.3. Fuzzy System with Fuzzy Input and Output

When A and B are fuzzy sets in X and Y respectively, the fuzzy relation $A \alpha B$ in $X \times Y$ is given by the following using the α operation:

$$A \circledast B \leftrightarrow \mu_{A \circledast B}(x, y) = \mu_A(x) \alpha \mu_B(y) \tag{3.21}$$

We thus obtain the following properties:

$$A \circ (A \circledast B) \subseteq B$$

$$R \subseteq A \circledast (A \circ R).$$

The following can be obtained from Equation (3.19) using these properties:

Property 3.1. Given fuzzy sets A and B, the largest R that satisfies fuzzy relational equation $A \circ R = B$ is the $\hat{R}$ below:

$$\hat{R} = A \circledast B \tag{3.22}$$

Example 3.10. If we find $A \circledast B$ for fuzzy sets $A = \dfrac{\begin{matrix} x_1 & x_2 & x_3 \end{matrix}}{[0.3 \quad 0.7 \quad 1]}$ and $B = \dfrac{\begin{matrix} y_1 & y_2 & y_3 \end{matrix}}{[1 \quad 0.6 \quad 0.4]}$ from Example 3.9, we get

$$\hat{R} = A \circledast B = \begin{matrix} \\ x_1 \\ x_2 \\ x_3 \end{matrix} \begin{matrix} y_1 & y_2 & y_3 \\ \begin{bmatrix} 1 & 1 & 1 \\ 1 & 0.6 & 0.4 \\ 1 & 0.6 & 0.4 \end{bmatrix} \end{matrix}.$$

This clearly satisfies $A \circ \hat{R} = B$, and we get $R \subseteq \hat{R}$ for the R from Example 3.9.

Note: For R that satisfies $A \circ \hat{R} = B$, we also have direct product $A \times B$ $(\leftrightarrow \mu_A(x) \wedge \mu_B(y))$. The above example comes out as follows, and we have $A \times B \subseteq \hat{R}$:

$$A \times B = \begin{matrix} \\ x_1 \\ x_2 \\ x_3 \end{matrix} \begin{matrix} y_1 & y_2 & y_3 \\ \begin{bmatrix} 0.3 & 0.3 & 0.3 \\ 0.7 & 0.6 & 0.4 \\ 1 & 0.6 & 0.4 \end{bmatrix} \end{matrix}.$$

Property 3.2. Given fuzzy relation R and fuzzy set B, the largest fuzzy set A that satisfies $A \circ R = B$ is

$$\hat{A} = R \circledast B, \tag{3.23}$$

where we have

$$\mu_{R \odot B}(x) = \bigwedge_y \{\mu_R(x, y) \, \alpha \, \mu_B(y)\},$$

which is guaranteed because of the following:

$$(R \otimes B) \circ R \subseteq B, \qquad A \subseteq R \otimes (A \circ R).$$

Example 3.11. Using the fuzzy relation R and fuzzy set B from Example 3.9, we get

$$
\hat{A} = R \otimes B =
\begin{array}{c}
 \\
x_1 \\
x_2 \\
x_3
\end{array}
\begin{array}{ccc}
y_1 & y_2 & y_3 \\
\left[\begin{array}{ccc}
0.5 & 0.7 & 0.8 \\
0.6 & 0.6 & 0.4 \\
1 & 0.5 & 0.3
\end{array}\right]
\end{array}
\otimes
\begin{array}{c}
\left[\begin{array}{c}
1 \\
0.6 \\
0.4
\end{array}\right]
\end{array}
\begin{array}{c}
y_1 \\
y_2 \\
y_3
\end{array}
$$

$$
= \begin{bmatrix}
1 \wedge 0.6 \wedge 0.4 \\
1 \wedge 1 \quad \wedge 1 \\
1 \wedge 1 \quad \wedge 1
\end{bmatrix}
= \begin{bmatrix}
0.4 \\
1 \\
1
\end{bmatrix}
\begin{array}{c}
x_1 \\
x_2 \\
x_3
\end{array}
$$

for $\hat{A}$, and it is clear that we can confirm that $\hat{A} \circ R = B, \quad A \subseteq \hat{A}$.

3.6 VARIOUS TYPES OF FUZZY RELATIONS

By adding restrictions to fuzzy relations, we can obtain various types of fuzzy relations. Examples of these are "similarity relations" and "fuzzy order relations." To place such limitations on fuzzy relations, we first need to take a look at the basic properties of those fuzzy relations.

Let R be a fuzzy relation in $X \times X$. For any $x, y, z \in X$, we have

(1) reflective: $\mu_R(x, x) = 1$;
(2) irreflexive: $\mu_R(x, x) = 0$;
(3) symmetric: $\mu_R(x, y) = \mu_R(y, x)$;
(4) antisymmetric: $\mu_R(x, y) > 0, \mu_R(y, x) > 0 \rightarrow x = y$,
   in other words, $x \neq y, \mu_R(x, y) > 0 \rightarrow \mu_R(y, x) = 0$;
(5) transitive: $\bigvee_y \{\mu_R(x, y) \wedge \mu_R(y, z)\} \leq \mu_R(x, z)$.

If we express these rules in terms of fuzzy operations, we come up with the following:

(1)′ $R \supseteq I$ (reflexive);
(2)′ $R \subseteq \bar{I}$ (irreflexive);
(3)′ $R = R^c$ (symmetrical);

$$\begin{bmatrix} 1 & 0.2 & 0 \\ 0.3 & 1 & 0.1 \\ 0.9 & 0.7 & 1 \end{bmatrix} \quad \begin{bmatrix} 0 & 0.5 & 0.4 \\ 0.3 & 0 & 0.2 \\ 1 & 0.9 & 0 \end{bmatrix} \quad \begin{bmatrix} 0.3 & 0.8 & 0.5 \\ 0.8 & 1 & 0.2 \\ 0.5 & 0.2 & 0 \end{bmatrix} \quad \begin{bmatrix} 0.3 & 0 & 0.5 \\ 0.8 & 1 & 0.7 \\ 0 & 0 & 0.1 \end{bmatrix} \quad \begin{bmatrix} 0.2 & 1 & 0.4 \\ 0 & 0.6 & 0.3 \\ 0 & 1 & 0.3 \end{bmatrix}$$

(a) Reflexive   (b) Inreflexive   (c) Symmetric   (d) Antisymmetric   (e) Transitive

Fig. 3.4.   Various Fuzzy Relations

(4)' $R \cap R^c \subseteq I$ (antisymmetrical);
(5)' $R \circ R \subseteq R$ (transitive).

Examples of fuzzy relations that satisfy these rules are shown in Fig. 3.4.
The transitivity in Fig. 3.4 (e) is clearly as follows:

$$R = \begin{bmatrix} 0.2 & 1 & 0.4 \\ 0 & 0.6 & 0.3 \\ 0 & 1 & 0.3 \end{bmatrix} \supseteq \begin{bmatrix} 0.2 & 0.6 & 0.3 \\ 0 & 0.6 & 0.3 \\ 0 & 0.6 & 0.3 \end{bmatrix} = R \circ R.$$

Example 3.12.   The fuzzy relation "x and y look alike" is reflexive and symmetrical; "x and y do not look alike" is irreflexive and symmetrical.

Example 3.13.   The fuzzy relation "$x \ll y$" is irreflexive, antisymmetrical, and transitive.

Fuzzy relations with the above limitations placed on them have the following properties.

Property 3.3.   In the case of reflexivity:

(1) If R is reflexive and S is any fuzzy relation, $R \circ S \supseteq S$ and $S \circ R \supseteq S$.
(2) If R is reflexive, $R \subseteq R \circ R$.
(3) If R and S are reflexive, $R \cup S$, $R \cap S$, and $R \circ S$ are also reflexive.

Property 3.4.   In the case of symmetry:

(1) If R and S are symmetrical, $R \cup S$, $R \cap S$, and R^m are also symmetrical.*
(2) If R and S are symmetrical and $R \circ S = S \circ R$, $R \circ S$ is also symmetrical.

Property 3.5.   In the case of transitivity:

(1) If R and S are transitive, $R \cap S$ is also transitive. $R \cup S$ is not always transitive.

* $R^m = \underbrace{R \circ R \circ \cdots \circ R}_{m}$

(2) If R and S are transitive and $R \circ S = S \circ R$, $R \circ S$ is transitive.
(3) If R is both symmetrical and transitive, $\mu_R(x, y) \leq \mu_R(x, x)$.
(4) If R is both reflexive and transitive, $R \circ R = R$.

Property 3.6. In the case of transitive closure:

(1) For any fuzzy relation R on X, when X has n number of elements, the transitive closure of R, $\hat{R}$, is transitive:

$$\hat{R} = R \cup R^2 \cup R^n. \tag{3.24}$$

(2) When R is reflexive ($R \supseteq I$),

$$I \subseteq R \subseteq R^2 \subseteq \cdots \subseteq R^{n-1} = R^n,$$

and therefore we have

$$\hat{R} = R^{n-1}.$$

(3) When R is transitive, we know $R \supseteq R^2 \supseteq R^3$ by definition, and $\hat{R}$ satisfies $\hat{R} = R$.
(4) When R is reflexive and transitive, we get $R = R^2 = R^3 = \cdots$, and $\hat{R}$ satisfies $\hat{R} = R$.
(5) For any fuzzy relation R,

$$(I \cup R)^m = I \cup R \cup R^2 \cup \cdots \cup R^m$$

$$\bigcup_{i=p}^{q} R^i = R^p \circ (I \cup R)^{q-p}.$$

If we let $p = 1$ and $q = n$, $\hat{R}$ can be expressed by

$$\hat{R} = R \cup R \cup \cdots \cup R^n = R \circ (I \cup R)^{n-1}.$$

3.7 SIMILARITY RELATIONS AND FUZZY ORDER RELATIONS

Let us now discuss similarity relations and fuzzy order relations as special cases of fuzzy relations. *Similarity relations* are extensions of ordinary equivalence relations, and we will explain the similarity classes and partition trees that are derived from them. *Fuzzy order relations* are standard order relations using a fuzzy approach, and we will give an explanation of fuzzy partial order relations and linear ordering.

$$
\begin{array}{c}
\begin{array}{ccccc} x_1 & x_2 & x_3 & x_4 & x_5 \end{array}\\
\begin{array}{c} x_1 \\ x_2 \\ x_3 \\ x_4 \\ x_5 \end{array}
\begin{bmatrix}
1 & 0.8 & 0.4 & 0.5 & 0.8\\
0.8 & 1 & 0.4 & 0.5 & 0.9\\
0.4 & 0.4 & 1 & 0.4 & 0.4\\
0.5 & 0.5 & 0.4 & 1 & 0.5\\
0.8 & 0.9 & 0.4 & 0.5 & 1
\end{bmatrix}
\end{array}
$$

Fig. 3.5. Similarity Relation S

3.7.1 Similarity Relations

When the following three conditions are met, the fuzzy relation S on X is a similarity relation.

(1) reflexivity: $\mu_S(x, x) = 1$; $S \supseteq I$;
(2) symmetry: $\mu_S(x, y) = \mu_S(y, x)$; $S = S^c$;
(3) transitivity: $\bigvee_y \{\mu_S(x, y) \wedge \mu_S(y, z)\} \leq \mu_S(x, z)$; $S \circ S \subseteq S$;

Similarity relations are sometimes called *fuzzy equivalence relations*.

Example 3.14. Let $X = \{x_1, x_2, \ldots, x_5\}$, and let S be the fuzzy relation in Fig. 3.5; S is a similarity relation.

Property 3.7. If S is a similarity relation on X, its α level-set S_α is an equivalence relation on X, where we have

$$
S_\alpha = \{(x, y) \mid \mu_S(x, y) \geq \alpha\}, \qquad 0 \leq \alpha \leq 1
$$

Example 3.15. We can find the α level-set for S from the previous example when, for example, $\alpha = 0.4, 0.5, 0.8, 0.9$, and 1. It is clear that these relations are equivalence relations:

$$
S_{0.4} = \begin{bmatrix}
1 & 1 & 1 & 1 & 1\\
1 & 1 & 1 & 1 & 1\\
1 & 1 & 1 & 1 & 1\\
1 & 1 & 1 & 1 & 1\\
1 & 1 & 1 & 1 & 1
\end{bmatrix}
\qquad
S_{0.5} = \begin{bmatrix}
1 & 1 & 0 & 1 & 1\\
1 & 1 & 0 & 1 & 1\\
0 & 0 & 1 & 0 & 0\\
1 & 1 & 0 & 1 & 1\\
1 & 1 & 0 & 1 & 1
\end{bmatrix}
$$

$$
S_{0.8} = \begin{bmatrix}
1 & 1 & 0 & 0 & 1\\
1 & 1 & 0 & 0 & 1\\
0 & 0 & 1 & 0 & 0\\
0 & 0 & 1 & 0 & 0\\
1 & 1 & 0 & 0 & 1
\end{bmatrix}
\qquad
S_{0.9} = \begin{bmatrix}
1 & 0 & 0 & 0 & 0\\
0 & 1 & 0 & 0 & 1\\
0 & 0 & 1 & 0 & 0\\
0 & 0 & 0 & 1 & 0\\
0 & 1 & 0 & 0 & 1
\end{bmatrix}
$$

$$S_1 = \begin{bmatrix} 1 & 0 & 0 & 0 & 0 \\ 0 & 1 & 0 & 0 & 0 \\ 0 & 0 & 1 & 0 & 0 \\ 0 & 0 & 0 & 1 & 0 \\ 0 & 0 & 0 & 0 & 1 \end{bmatrix}$$

The equivalence classes for these equivalence relations are

$$\{\{x_1\}, \{x_2\}, \{x_5\}, \{x_4\}, \{x_3\}\} \cdots S_1 \text{ in case of } S_1$$

$$\{\{x_1\}, \{x_2, x_5\}, \{x_4\}, \{x_3\}\} \quad \cdots S_{0.9} \text{ in case of } S_{0.9}$$

$$\{\{x_1, x_2, x_5\}, \{x_4\}, \{x_3\}\} \quad \cdots S_{0.8} \text{ in case of } S_{0.8}$$

$$\{\{x_1, x_2, x_5, x_4\}, \{x_3\}\} \quad \cdots S_{0.5} \text{ in case of } S_{0.5}$$

$$\{\{x_1, x_2, x_5, x_4, x_3\}\} \quad \cdots S_{0.4} \text{ in case of } S_{0.4}.$$

These elements can be summed up in a tree-like structure, as in Fig. 3.6. We can see that as α gets larger, the partitions get smaller. For this reason, the tree in Fig. 3.6 is called a *partition tree*.

3.7.2   Similarity Classes

The concept of finding an equivalence class by means of an equivalence relation can be applied to similarity relations. If we let S be a similarity relation

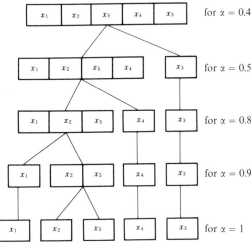

Fig. 3.6.   Partition Tree for Similarity Relation S

for $X = \{x_1, x_2, \ldots, x_n\}$, the *similarity class* for x_i, $[x_i]_S$ is a fuzzy set in X, and its membership function is

$$\mu[x_i]_S(x_j) = \mu_S(x_i, x_j).$$

The similarity class can be expressed in terms of the equivalence class $[x_i]_{S_\alpha}$ of the level-set S_α equivalence relation;

$$[x_i]_S = \bigcup_\alpha \alpha[x_i]_S.$$

The equivalence classes for ordinary equivalence relations are disjoint, but similarity classes by means of similarity relations are not necessarily so. The degree to which similarity classes $[x_i]_S$ and $[x_j]_S$ are not disjoint is defined as the height of the intersection of $[x_i]_S$ and $[x_j]_S$, and this gives rise to the following relation:

$$\text{hgt}([x_i]_S \cap [x_j]_S) = \mu_S(x_i, x_j).$$

Example 3.16.   The similarity classes $[x_1]_S$ and $[x_2]_S$, obtained by means of the similarity relations in Fig. 3.5, are

$$[x_1]_S = 1/x_1 + 0.8/x_2 + 0.4/x_3 + 0.5/x_4 + 0.8/x_5$$

$$[x_2]_S = 0.8/x_1 + 1/x_2 + 0.4/x_3 + 0.5/x_4 + 0.9/x_5,$$

and the extent to which $[x_1]_S$ and $[x_2]_S$ are not disjoint is

$$\text{hgt}([x_1]_S \cap [x_2]_S)$$

$$= \text{hgt}(0.8/x_1 + 0.8/x_2 + 0.4/x_3 + 0.5/x_4 + 0.8/x_5)$$

$$= 0.8$$

$$= \mu_S(x_1, x_2).$$

3.7.3   Resemblance Relations

The fuzzy relation "look alike" is reflexive and symmetrical but not transitive, so it is not a similarity relation. However, this is a relation that we frequently come across, so we have the following "resemblance relation" that corresponds to it.

For fuzzy relation R to be a resemblance relation, it has to satisfy the following two conditions:

(1) reflexivity: $\mu_R(x, x) = 1$;
(2) symmetry: $\mu_R(x, y) = \mu_R(y, x)$;

The transitive closure for any fuzzy relation is transitive (Property 3.6), so the following can be said.

Property 3.8. The transitive closure $\hat{R}$ of any resemblance relation $\hat{R}$ is a similarity relation

Next we will discuss fuzzy partial order relations and fuzzy linear ordering as generalizations of their standard counterparts.

3.7.4 Fuzzy Partial Order Relations

For fuzzy relation P to be a *fuzzy partial order relation* on X, the following conditions must be satisfied:

(1) reflexivity: $\mu_P(x, x) = 1$;
(2) antisymmetry: $x \neq y$, $\mu_P(x, y) > 0 \rightarrow \mu_P(y, x) = 0$;
(3) transitivity: $\bigvee_y \{\mu_P(x, y) \wedge \mu_P(y, z)\} \leq \mu_P(x, z)$.

Example 3.17. The fuzzy relation in Fig. 3.7(a) is a fuzzy partial order relation. Hasse diagram for it is shown in Fig. 3.7(b).

In the case of a similarity relation, the level-set is an equivalence relation (Property 3.7), but what happens in the case of fuzzy partial order relations?

Property 3.9. Let P be a fuzzy partial order relation on X. Its α level-set P_α is a partial order relation on X.

Now let us define upper bounds, lower bounds, least upper bounds, and greatest upper bounds for fuzzy order relations, all of which are important concepts for standard order relations.

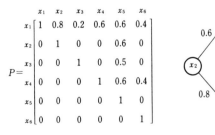

$$P = \begin{array}{c c} & \begin{array}{cccccc} x_1 & x_2 & x_3 & x_4 & x_5 & x_6 \end{array} \\ \begin{array}{c} x_1 \\ x_2 \\ x_3 \\ x_4 \\ x_5 \\ x_6 \end{array} & \begin{bmatrix} 1 & 0.8 & 0.2 & 0.6 & 0.6 & 0.4 \\ 0 & 1 & 0 & 0 & 0.6 & 0 \\ 0 & 0 & 1 & 0 & 0.5 & 0 \\ 0 & 0 & 0 & 1 & 0.6 & 0.4 \\ 0 & 0 & 0 & 0 & 1 & 0 \\ 0 & 0 & 0 & 0 & 0 & 1 \end{bmatrix} \end{array}$$

(a) Fuzzy Order Relation P (b) Hasse (E) Diagram

Fig. 3.7. Fuzzy Order Relation and Hasse (E) Diagram

3.7.5   Dominating Class and Dominated Class

The *dominating class* for element x_i of X is expressed by $P \geq [x_i]$ and is a fuzzy set in X defined by

$$\mu_{P \geq [x_i]}(x_j) = \mu_P(x_i, x_j).$$

In the same way, the *dominated class* for $x_i P \leq [x_i]$ is defined as follows:

$$\mu_{P \leq [x_i]} = \mu_P(x_j, x_i).$$

For x_i to be the *maximal element*, we have

$$\mu_P(x_i, x_j) = 0; \qquad {}^\forall x_j \neq x_i$$

for all $x_j (\neq x_i)$. In the same way, we have the *minimal element* when

$$\mu_P(x_j, x_i) = 0; \qquad {}^\forall x_j \neq x_i.$$

Example 3.18.   For fuzzy partial order relation P in Fig. 3.7, x_5 and x_6 are maximal elements, and x_1 is a minimal element.

3.7.6 Fuzzy Upper and Lower Bounds

If A is a subset of X, the *fuzzy upper bound* is represented as $U(A)$ and is defined as

$$U(A) = \bigcap_{x_i \in A} P \geq [x_i].$$

In the same way, the *fuzzy lower bound* of A, $L(A)$, is defined as

$$L(A) = \bigcap_{x_i \in A} P \leq [x_i].$$

Example 3.19.   Let $A = \{x_2, x_3, x_4\}$ in Fig. 3.7. The fuzzy upper bound $U(A)$ and fuzzy lower bound $L(A)$ for A are

$$\begin{aligned}
U(A) &= P \geq [x_2] \cap P \geq [x_3] \cap P \geq [x_4] \\
&= (1/x_2 + 0.6/x_5) \cap (1/x_3 + 0.5/x_5) \cap (1/x_4 + 0.6/x_5 + 0.4/x_6) \\
&= 0.5/x_5
\end{aligned}$$

$$\begin{aligned}
L(A) &= P \leq [x_2] \cap P \leq [x_3] \cap P \leq [x_4] \\
&= 0.2/x_1.
\end{aligned}$$

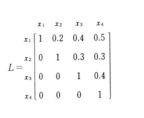
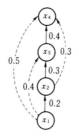

(a) Fuzzy Linear Ordering L (b) Hasse (E) Diagram

Fig. 3.8. Fuzzy Linear Ordering and Hasse (E) Diagram

3.7.7 Least Upper Bounds and Greatest Lower Bounds

If it exists, the *least upper bound* of A is an x_i that satisfies

$$\mu_{U(A)}(x_i) > 0, \qquad \mu_P(x_i, x_j) > 0; \qquad {}^{\forall}x_j \in \text{supp}(U(A)).*$$

In other words the least upper bound of A is the smallest element of $U(A)$.

In the same way, if it exists, the *greatest lower bound* of A is an x_i such that

$$\mu_{L(A)}(x_i) > 0, \qquad \mu_P(x_j, x_i) > 0; \qquad {}^{\forall}x_j \in \text{supp}(L(A)).$$

In other words, the least upper bound of A is the largest element of $L(A)$.

3.7.8 Special Fuzzy Order Relations

Finally, we will talk about some other special fuzzy order relations.

Fuzzy Linear Ordering

For fuzzy relation L in X to be *fuzzy linear ordering*, the following conditions must be satisfied:

(1) it must be a fuzzy partial order relation;
(2) it must show comparability:

$$x_i \neq x_j \rightarrow \mu_L(x_i, x_j) > 0 \quad \text{or} \quad \mu_L(x_j, x_i) > 0.$$

Example 3.21.   Fig. 3.8 shows fuzzy linear ordering. Also, the fuzzy relation "$\ll$" (much greater than) is an irreflexive fuzzy linear ordering.

---

* The *support* of fuzzy set A, expressed supp(A), is a subset of X such that supp(A) = $\{x \mid \mu_A(x) > 0\}$

Table 3.4. Various Fuzzy Relations

Fuzzy Relation \ Condition	Reflexivity	Irreflexivity	Symmetry	Antisymmetry	Transitivity	Comparability
Similarity Relation	○		○		○	
Resemblance Relation	○		○			
Fuzzy Partial Order Relation	○			○	○	
Fuzzy Linear Ordering	○			○	○	○
Fuzzy Preorder Relation	○				○	
Fuzzy Weak Order Relation	○				○	○

Fuzzy Preorder Relations

A *fuzzy preorder relation* on X must satisfy the following conditions:

(1) reflexivity: $\mu_R(x, x) = 1$;
(2) transitivity: $\bigvee_y \{\mu_R(x, y) \wedge \mu_R(y, z)\} \leq \mu_R(x, z)$.

From this it is clear that fuzzy partial order relations and similarity relations are fuzzy preorder relations.

Now that we have added various conditions to fuzzy relations to get similarity relations, fuzzy order relations, etc., let us sum them alll up in a table (Table 3.4).

REFERENCES

(1) Zadeh, L. A., "Similarity Relations and Fuzzy Ordering," *Information Sciences* **3**, pp. 177–200 (1971).
(2) Zadeh, L. A., "Calculus of Fuzzy Restriction," in Zadeh, Tanaka, *et al.*, eds., *Fuzzy Sets and Their Applications to Cognitive and Decision Processes*, pp.1–39, Academic Press, New York. (1975).
(3) Mamdani, E. H., "Application of Fuzzy Logic to Approximate Reasoning Using Linguistic Systems," *IEEE Transactions on Computer*, **C**–26, pp. 1182–1191 (1977).
(4) Mizumoto, M., Comparison of Various "Fuzzy Reasoning Methods," *Preprints of Second IFSA Congress*, pp. 2–7 (1987).
(5) Mizumoto, M., "Extended Fuzzy Reasoning," in Gupta, *et al.*, eds., *Approximate Reasoning in Expert Systems*, pp. 71–85, Elsevier Science Pub. (1985).
(6) Sanchez, E., "Resolution of Composite Fuzzy Relation Equations," *Information and Control*, **30**, pp. 38–47 (1976).

Chapter 4

FUZZY REGRESSION MODELS

With standard regression models, the difference between the data and the inferred value obtained from the model is taken to be observational error, but with fuzzy regression models, it is assumed that the gap between the data and the model is an ambiguity in the structure of the system that gives the input and output. Here, we will consider the coefficient that expresses the system's ambiguity, and we will deal with systems for which coefficients are expressed by fuzzy numbers. Since the fuzzy number for the coefficient gives the possibilities for the coefficient, the system is called a linear possibility system. The model that expresses input and output data by means of a possibility system is called a *fuzzy regression model*, and we will explain a new type of data analysis that brings the ambiguity of this kind of input/output relation back to the system coefficient.

4.1 LINEAR POSSIBILITY SYSTEMS

One explanation for fuzzy sets is that the membership function can be seen as a possibility distribution. The information from a specialist, expressed by something like "about 10" $\triangleq F$, is replaced by $\mu_F(x)$, and this new expression is taken as the information that expresses possibility. For example, if $\mu_F(8) = 0.8$, the degree of possibility of 8 is 0.8. In order to make positive use of this kind of information, we look at the given information F in terms of possibility using the possibility distribution function $\mu_F(x)$.

As in the case of probability, if the possibility function $\mu_F(x)$ is given, we can think about learning the possibility of fuzzy event A by means of the fuzzy set. Easily understood possibility functions are given in terms of intervals, and we can consider a case in which the event is also an interval. Let $F = \{x \mid 0 \leq x \leq 5\}$ and $A = \{x \mid 2 \leq x \leq 7\}$. We can think of the possibility of event A based on the possibility of information F as being 1. This is because $F \cap A \neq \phi$. If $A' = \{x \mid 6 \leq x \leq 10\}$, the possibility of A' is 0. If this idea is extended to fuzzy sets, we have the following definition.

Definition 4.1. *When possibility function $\mu_F(x)$ is given, the possibility measure $\pi(A)$ is defined as*

$$\pi(A) = \sup_x \mu_A(x) \wedge \mu_F(x) \tag{4.1}$$

Ambiguous information such as "about 10" is given by a fuzzy number. Fuzzy numbers have already been discussed in Chapter 2. The information obtained from the fuzzy numbers is given as a possibility distribution function, and making use of Definition 4.1, we obtain the following properties:

(1) $\pi_X(\phi) = 0, \qquad \pi_X(X) = 1$

(2) $\pi_X(A \cup B) = \pi_X(A) \vee \pi_X(B)$.

As is clear from (2), the possibility measure is assumed to have the following monotony only:

$$A \subset B \rightarrow \pi_X(A) \leq \pi_X(B). \tag{4.2}$$

As in the case of probability, where various types of probability distribution functions have been considered, there are many kinds of possibility distribution functions that can be considered, but in this chapter we will deal only with L–R fuzzy numbers.[1]

Definition 4.2. *Fuzzy number A is expressed $A = (\alpha, c)$, and its membership function is expressed as follows:*

$$\mu_A(x) = L((x - \alpha)/c); \qquad c > 0. \tag{4.3}$$

where $L(x)$ is called a reference function and (1) $L(x) = L(-x)$, (2) $L(0) = 1$, and (3) $L(x)$ is a strictly decreasing function for $[0, \infty)$. In addition, α expresses the center and c the spread or width.

Examples of $L(x)$ for $p > 0$ are functions like $L_1(x) = \max(0, 1 - |x|^p)$, $L_2(x) = e^{-|x|^p}$, $L_3(x) = 1/(1 + |x|^p)$. If we consider $L_1(x)$ for $p = 1$, we get the triangular fuzzy number in Fig. 4.1 for fuzzy number A in (4.3). In the figure, α is the center and c the width.

Let us now define the operations with fuzzy numbers by means of the extension principle.[2]

Definition 4.3. *When the function $y = f(x_1, \ldots, x_n)$ is given and the input $x_1, \ldots, x_n$ is replaced by $A_1, \ldots, A_n$, the fuzzy output is defined by*

$$\mu_Y(y) = \sup_{\{x_1, \ldots, x_n | y = f(x_1, \ldots, x_n)\}} \mu_{A_1}(x_1) \wedge \cdots \wedge \mu_{A_n}(x_n) \qquad (4.4)$$

In other words, this means finding x on the set $\{x_1, \ldots, x_n | y = f(x_1, \ldots, x_n)\}$ such that it maximizes $\mu_{A_1}(x_1) \wedge \cdots \wedge \mu_{A_1}(x_n)$, after y is fixed.

This is called the *extension principle*, because the function for x is extended to a function of fuzzy numbers. Operations for symmetrical fuzzy numbers are as follows:

(1) addition: $(\alpha_1, c_1)_L + (\alpha_2, c_2)_L = (\alpha_1 + \alpha_2, c_1 + c_2)_L;$ (4.5.1)
(2) subtraction: $(\alpha_1, c_1)_L - (\alpha_2, c_2)_L = (\alpha_1 - \alpha_2, c_1 + c_2)_L;$ (4.5.2)
(3) scalar multiplication: $\lambda \circ (\alpha, c) = (\lambda\alpha, |\lambda|_c)_L.$ (4.5.3)

If we make $L(x)$ triangular, the fact that (4.5.1–3) hold can easily be shown. We will express the linear possibility system using symmetrical fuzzy number $A_i = (\alpha_i, c_i)_L$ as

$$Y = A_1 x_1 + \cdots + A_n x_n \qquad (4.6)$$

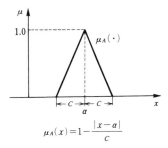

$$\mu_A(x) = 1 - \frac{|x - \alpha|}{c}$$

Fig. 4.1. Example of a Triangular Fuzzy Number: $\mu_A(x) = 1 - \dfrac{|x - \alpha|}{c}$

The membership function for fuzzy number A_i is

$$\mu_{A_i}(a_i) = L((a_i - \alpha_i)/c_i) \tag{4.7}$$

Theorem 4.1. *The membership function for the output of a linear possibility system is as follows:*

$$\mu_Y(y) = L((y - \mathbf{x}^t\boldsymbol{\alpha})/\mathbf{c}^t|\mathbf{x}|), \tag{4.8}$$

where $|\mathbf{x}| = (|x_1|, \ldots, |x_n|)^t$.

Proof is given in Reference (8). If the possibility distribution $(\alpha_i, c_i)_L$ for the coefficient in (4.6) is given, the possibility distribution for output y can be calculated from Theorem 4.1. In other words, the center for possibility distribution Y is $\Sigma\alpha_i x_i$ and the width is $\Sigma c_i|x_i|$.

In order to make a comparison with probability, let us look at the following. When the coefficient is the possibility distribution that is fuzzy number $A_j = (\alpha_j, c_j)_L$, it can be expressed symbolically as follows for comparison with the independent probability distribution in which the coefficient turns out to be a normal distribution $N(e_j, \sigma_j^2)$:

possibility distribution $(\alpha_1, c_1)_L x_t + \cdots + (\alpha_n, c_n)_L x_n = (\boldsymbol{\alpha}^t\mathbf{x}, \mathbf{c}^t|\mathbf{x}|)_L$

probability distribution $N(e_1, \sigma_1^2)x_1 + \cdots + N(e_n, \sigma_n^2)x_n \tag{4.9}$

$$= N(\mathbf{e}^t\mathbf{x}, (\boldsymbol{\sigma}^2)^t\mathbf{x}^2),$$

where $\mathbf{x}^2 = (x_1^2, \ldots, x_n^2)^t$.

From the above, we can see that possibility is calculated based on possibility measures and corresponds to probability calculated on the basis of probability measures. The standpoint from which ambiguity is handled differs.

Finally, let us define the inclusion relation for fuzzy numbers.

Definition 4.4. *The inclusion relation for symmetrical fuzzy numbers with a degree h is defined as follows:*

$$A \underset{h}{\supseteq} B \leftrightarrow [A]_h \supset [B]_h. \tag{4.10}$$

4.2 LINEAR POSSIBILITY REGRESSION MODEL

Here we will discuss the linear regression method using the linear possibility system discussed in 4.1. We will divide our explanation according to the type of data handled: standard data and data for which the output is fuzzy numbers.

4.2.1   Standard Data

The given data are $(y_i, x_{i1}, \ldots, x_{in})$, $i = 1, \ldots, N$. y_i is an output for the ith sample and x_{ij} is the jth input or jth explanatory variable for the ith sample. The vector for the explanatory variables is expressed as $x_i = (x_{i1}, \ldots, x_{in})^t$. With standard regression models, the difference between the actual data and the inferred values,

$$y_j - \sum a_i x_{ji} = \varepsilon_j; \qquad j = 1, \cdots, N \tag{4.11}$$

is interpreted as observational error, and the regression analysis is performed by means of the probability model. The linear possibility regression analysis that we are talking about here follows the possibility model. The uncertainty in the data is seen as originating in the system itself, and it depends on the possibility of the system coefficient. Since the formulation that gives us the linear possibility regression model is performed from this position, we establish the following.

(1) We let the linear possibility system be the model, that is,

$$Y_i = A_0 + A_1 x_{i1} + \cdots + A_n x_{in}, \tag{4.12}$$

where fuzzy coefficient A_i is symmetrical fuzzy number $(\alpha_i, c_i)_L$.

(2) Fuzzy coefficient A_i is determined by the degree h to which the given data (y_i, x_i) is included in the inferred fuzzy number Y_i. More precisely,

$$\mu_{Y_i}(y_i) \geq h; \qquad i = 1, \ldots, N, \tag{4.13}$$

where Y_i is the inferred fuzzy number from (4.12).

(3) The fuzzy coefficient A_i that minimizes the total width of Y_i is determined, that is,

$$J(c) = \sum c^t |x_i|, \tag{4.14}$$

where $c = (c_1, \ldots, c_n)^t$, and $c^t |x_i|$ is the width of Y_i. $J(c)$ corresponds to the sum of errors in conventional regression analysis.

The above is shown in Fig. 4.2.

Based on what has been established above, the linear possibility regression problem is reduced to finding the $A_i = (a_i, c_i)_L$ that minimizes the objective function (4.14) subject to the constraint (4.13). This brings us back to the

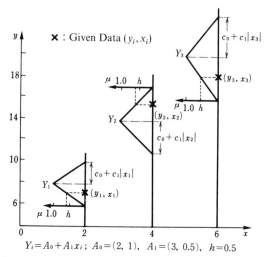

Fig. 4.2. Linear Possibility Regression Model: $Y_i = A_0 + A_1 x_i$; $A_0 = (2, 1)$, $A_1 = (3, 0.5)$, $h = 0.5$

following linear programming (LP) problem:

$$
\left.
\begin{aligned}
\min_{\alpha, c} J(c) &= \sum c^t |x_i| \\
y_i &\leq x_i^t \alpha + |L^{-1}(h)| c^t |x_i| \\
y_i &\geq x_i^t \alpha - |L^{-1}(h)| c^t |x_i| \\
c &\geq 0, \qquad i = 1, \ldots, N
\end{aligned}
\right\}.
\tag{4.15}
$$

In LP problem (4.15), (4.13) can be replaced by the above constraint, due to Theorem 4.1. The linear possibility model can be obtained by solving the LP problem.

Theorem 4.2 *Given data (y_i, x_i), there is an optimal solution $A^* = (\alpha^*, c^*)_L$ for $0 \leq h < 1$ in LP problem (4.15).*

If c_i is large enough, it is easy to see that an admissible set exists for LP problem (4.15). When $h = 1$, the data must satisfy

$$
y_i = x_i^t \alpha; \qquad i = 1, \ldots, N.
\tag{4.16}
$$

(4.16) is not generally given rise to, so $h = 1$ is excluded. Theorem 4.2 says that the linear possibility system that includes given data (y_i, x_i) to degree h already exists.

Now let the optimal h level solution be $A_h = (\alpha_h, c_h)_L$.

Theorem 4.3. *The optimal solution for $h' \neq h$ can be obtained from the optimal h level solution in the following way:*

$$A_{h'} = \left(\alpha_h, \frac{L^{-1}(h)}{L^{-1}(h')} c_h \right)_L. \tag{4.17}$$

From Theorem 4.3, we can see that after we find the $h = 0$ level solution, we can easily obtain the optimal solution for any level. Also, the value of J for any h is

$$J(c_h) = \frac{|L^{-1}(0)|}{|L^{-1}(h)|} J(c_0), \tag{4.18}$$

and if we let $L(x) = 1 - |x|$, we get

$$J(c_h) = (1 - h)^{-1} J(c_0). \tag{4.19}$$

From (4.19), the relationship between, for example, solution c_0 obtained at $h = 0$ and solution $c_{0.5}$ is

$$2J(c_0) = J(c_{0.5}) \tag{4.20}$$

From the above, degree h can be interpreted in the following manner. If there are enough data, the solution for $h = 0$ is found. In other words, since the data include all of the possibilities, it is sufficient to consider only those possibilities. If the data are considered to include only half, $h = 0.5$ is used. Since this means that a width twice the width obtained from the data is obtained, we think of it as adding the width of the obtainable data to the width of the unobtainable data.

4.2.2 Fuzzy Data

$(Y_i, x_i), i = 1, \dots, N$ are the data we will use here. Y_i expresses the fuzzy output and is expressed $Y_i = (y_i, e_i)_L$. Given the fuzzy data (Y_i, x_i), the basic idea is to find $\underline{A}$ and $\bar{A}$ given by

$$\underline{Y}_i = \underline{A}_1 x_{i1} + \cdots + \underline{A}_n x_{in} \underset{h}{\subseteq} Y_i \underset{h}{\subseteq} \bar{A}_1 x_{i1} + \cdots + \bar{A}_n x_{in} = \bar{Y}_i. \tag{4.21}$$

(4.21) leads us to a consideration of the following two problems.

Min Problem

$$
\left.
\begin{aligned}
\min_{\bar{A}_i = (\bar{a}_i, \bar{c}_i)_L} \quad & J(\bar{c}) = \sum \bar{c}^t |x_i| \\
& \bar{Y}_i = \bar{A}_1 x_{i1} + \cdots + \bar{A}_n x_{in} \underset{h}{\supseteq} Y_i; \qquad i = 1, \ldots, N
\end{aligned}
\right\}. \tag{4.22}
$$

This problem allows us to find $\bar{A}$ in the same way that we find the inferred fuzzy output $[\bar{Y}_j]_h$, which includes fuzzy output $[\bar{Y}_j]_h$. In other words, we find an inferred fuzzy output $\bar{Y}_i$ that minimizes the total width of $\bar{Y}_i$, based on the constraint $[Y_j]_h \subset [\bar{Y}_j]_h$. Min problem (4.22) returns us to the following LP problem:

$$
\left.
\begin{aligned}
\min_{\bar{\alpha}, \bar{c}} & \sum_i \bar{c}^t |x_i| \\
& y_i + |L^{-1}(h)| e_i \le x_i^t \bar{\alpha} + |L^{-1}(h)| \bar{c}^t |x_i| \\
& y_i - |L^{-1}(h)| e_i \ge x_i^t \bar{\alpha} - |L^{-1}(h)| \bar{c}^t |x_i|
\end{aligned}
\right\}. \tag{4.23}
$$

Max Problem

$$
\left.
\begin{aligned}
\max_{\underline{A}_h = (\alpha_h, c_h)_L} \quad & J(\underline{c}) = \sum \underline{c}^t |x_i| \\
& \underline{Y}_i = \underline{A}_1 x_{i1} + \cdots + \underline{A}_n x_{in} \underset{h}{\subseteq} Y_i; \qquad i = 1, \ldots, N
\end{aligned}
\right\}. \tag{4.24}
$$

This problem allows us to find fuzzy coefficient $\underline{A}$ in the same way that we find inferred fuzzy output $[\underline{Y}_i]_h$, which is included in fuzzy output $[Y_i]_h$. In other words, we find the fuzzy coefficient $\underline{A}_i$ that maximizes the total width of $[Y_i]_h$, based on the constraint $[\underline{Y}_i]_h \subset [Y_i]_h$. The max problem (4.24) turns out to be the following LP problem:

$$
\left.
\begin{aligned}
\max_{\alpha, c} & \sum_i \underline{c}^t |x_i| \\
& y_i + |L^{-1}(h)| e_i \ge x_i^t \underline{\alpha} + |L^{-1}(h)| \underline{c}^t |x_i| \\
& y_i - |L^{-1}(h)| e_i \le x_i^t \underline{\alpha} - |L^{-1}(h)| \underline{c}^t |x_i|
\end{aligned}
\right\}. \tag{4.25}
$$

For the min problem in (4.22), fuzzy output $\bar{Y}_i$ is inferred from the top down, and the inference is carried out so as to minimize the width of $\bar{Y}_j$. In the same way, for the max problem in (4.24), fuzzy output $\underline{Y}_j$ is inferred from the bottom up, and the inference is done so as to maximize the width of $\underline{Y}_j$.

Theorem 4.4. *If the given data* (Y_i°, x_i°) *satisfy* $Y_i^\circ = A^\circ x_i$, *the solutions to the min and max problems are*

$$
A^\circ = \bar{A} = A, \qquad Y^\circ = \bar{Y} = Y. \tag{4.26}
$$

If the given data (Y_i°, x_i°) satisfy the linear possibility system, it means that the solutions to the min and max problems identify fuzzy parameters, and it shows that the inferred output Y agrees with the given fuzzy output data Y°.

Since actual input and output data (Y_i°, x_i°) do not always satisfy the linear possibility system, these min and max solutions must be examined.

Theorem 4.5. *A solution for the min problem already exists, but one for the max problem cannot be guaranteed.*

If c in the min problem (4.23) is large enough, the constraint conditions are satisfied regardless of the data (Y_i, x_i), so the solution already exists. However, in the max problem (4.25) conditions are not necessarily satisfied even if $c = 0$, so a solution cannot be guaranteed.

Theorem 4.6. *If there is a solution to the max problem, we have*

$$Y_i \underset{h}{\subseteq} Y_i \underset{h}{\subseteq} \bar{Y}_i. \tag{4.27}$$

This is clear from the structure of the min and max problems. The solution to the max problem is called the least upper bound of the fuzzy estimate and the solution to the min problem the greatest lower bound. Given fuzzy data, a fuzzy inference can be obtained as to the greatest lower bound and the least upper bound.

4.3 EXAMPLES OF APPLICATIONS

We will divide the applications below, as we did in 4.2, into those with standard data and those with fuzzy data.

4.3.1 Regression Analysis with Standard Data

Here we will generate a linear possibility regression model for evaluating the cost of houses built by prefabricated housing Company A using their 1978 pamphlet. The data obtained from the pamphlet are as follows.

Input data: x_1 = quality of materials, x_2 = 1st floor area, x_3 = second floor area, x_4 = total number of rooms, x_5 = number of Japanese rooms

Output data: y = selling price of house [10K yen]

Table 4.1. Input/Output Data for Prefab Houses

No.	y_i	x_1	x_2	x_3	x_4	x_5
1	606	1	38.09	36.43	5	1
2	710	1	62.10	26.50	6	1
3	808	1	63.76	44.71	7	1
4	826	1	74.52	38.09	8	1
5	865	1	75.38	41.40	7	2
6	852	2	52.99	26.49	4	2
7	917	2	62.93	26.49	5	2
8	1 031	2	72.04	33.12	6	3
9	1 092	2	76.12	43.06	7	2
10	1 203	2	90.26	42.64	7	2
11	1 394	3	85.70	31.33	6	3
12	1 420	3	95.27	27.64	6	3
13	1 601	3	105.98	27.64	6	3
14	1 632	3	79.25	66.81	6	3
15	1 699	3	120.50	32.25	6	3

The actual data are shown in Table 4.1. The materials are: $1 =$ low quality, $2 =$ average quality, $3 =$ high quality. Using this data, let us assume that

$$Y_i = A_0 + A_1 x_{i1} + \cdots + A_5 x_{i5} \tag{4.28}$$

is the linear possibility function that gives the selling price of Company A's prefabricated houses. A_0 is a fuzzy constant. If, for simplicity, we let the fuzzy coefficient be triangular, that is $L(x) = 1 - |x|$, we get the problem for finding fuzzy coefficient A from (4.15):

$$\left. \begin{aligned} &\min_{\alpha,c} J(c) = \sum c^t |x_i| \\ &y_i \leq x_i^t \alpha + (1 - h) c^t |x_i| \\ &y_i \geq x_i^t \alpha - (1 - h) c^t |x_i|, \quad \alpha, c \geq 0 \end{aligned} \right\} . \tag{4.29}$$

The result of the LP problem for $h = 0.5$ is shown in the following:

$$\left. \begin{aligned} A_0^* &= (0,0)_L, & A_1^* &= (245.17, 37.63)_L \\ A_2^* &= (5.85, 0)_L, & A_3^* &= (4.79, 0)_L \\ A_4^* &= (0,0)_L, & A_5^* &= (0,0)_L \end{aligned} \right\} . \tag{4.30}$$

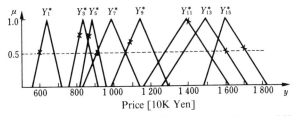

Fig. 4.3.   Inferred Fuzzy Output $Y_i^* = A^* x_i$ ($\times$: Observed Value)

The inferred fuzzy values for $Y_i^* = A^* x_i$ obtained from the odd-numbered data are also shown in Fig. 4.3. Since the problem was solved using $h = 0.5$, $\mu_{Y_i}(y_i) \geqq 0.5$. Actual values for $\mu_{Y_i}(y_i)$ are shown in Table 4.2. $[Y_i^*]_0 = \{y \mid \mu_{Y_i^*}(y_i) \geqq 0\}$ is shown in Table 4.3. We can consider the possible price of prefabricated house #1 as ranging from ¥5,671,900 to ¥7,177,300, and explain the actual ¥6,070,000 as having been chosen for the actual price. Since A_0, A_4 and A_5 are $(0, 0)_L$, the constants, total number of rooms, and number of Japanese rooms are not chosen to be variables. Fig. 4.3 also shows the relationship between the observed values ($\times$) and the inferred values. From Table 4.2, we see that samples 4, 6, 13, and 14 for which $\mu_{Y_i}(y_i) = 0.500$ are endpoints.

Table 4.2.   Possibility $\mu_{Y_i}^*(y_i)$
for Inferred Fuzzy Number Y_i^*
for Data y_i

No.	$\mu_{Y_i}^*(y_i)$	No.	$\mu_{Y_i}^*(y_i)$
1	0.516	9	0.668
2	0.662	10	0.869
3	0.677	11	0.969
4	0.500	12	0.976
5	0.741	13	0.500
6	0.500	14	0.500
7	0.545	15	0.540
8	0.738		

Table 4.3. Lower Bounds, Centers, Upper Bounds of the
0-Level Sets for Inferred Fuzzy Output Y_i^*

No.	Data y_i	Lower Bound	Center	Upper Bound
1	606	567.194	642.462	717.730
2	710	660.199	735.467	810.735
3	808	757.068	832.336	907.604
4	826	788.363	863.631	938.899
5	865	809.239	884.507	959.775
6	852	776.730	927.266	1 077.800
7	917	834.908	985.444	1 135.980
8	1 031	919.960	1 070.500	1 221.030
9	1 092	991.414	1 141.950	1 292.490
10	1 203	1 072.160	1 222.700	1 373.240
11	1 394	1 161.240	1 387.050	1 612.850
12	1 420	1 199.600	1 425.400	1 651.210
13	1 601	1 262.280	1 488.090	1 713.890
14	1 632	1 293.300	1 519.100	1 744.910
15	1 699	1 369.330	1 595.140	1 820.940

4.3.2 Regression Analysis with Fuzzy Data

The relationship between the value of the yen and the trading conditions for businesses is ambiguous in a very complicated way. Therefore, it can be thought more appropriate to try and grasp the changes in the value of the yen in terms of possibility rather than probability. Here we will identify a linear possibility system in which the input x_i is the trading conditions and the output Y_i is the value of the yen.

Oil, chemical, iron and steel, electric machine, and transport machine industries are considered as the input data for "type of enterprise," and adding in the constant, there are six variables. Output Y_i is given in terms of center y_i and width e_i. A numerical example of trading conditions and yen value is shown in Table 4.4. Actually, we let

$$\text{Trading Conditions} = \frac{\text{Production Price Index}}{\text{Investment Price Index}},$$

and is an index formed by assigning the year 1975 a value of 100. Using samples 1–13, we find the fuzzy coefficients for

$$Y_i = A_0 + A_1 x_{i1} + \cdots + A_5 x_{i5}, \tag{4.31}$$

Table 4.4. Numerical Example of Business Conditions and Yen Rate

Data for Structural Identification

Sample	Time (Quarter)	x_1	x_2	x_3	x_4	x_5	$Y_i = (y_i, e_i)$ (¥)
	1979						
1	III	106.1	89.3	111.6	83.3	89.1	(218.75, 10.94)
2	IV	98.1	84.1	104.4	78.0	88.8	(238.37, 11.92)
	1980						
3	I	89.2	76.7	100.1	70.8	87.5	(243.38, 12.17)
4	II	96.8	71.1	95.9	71.3	81.7	(233.20, 11.66)
5	III	101.7	68.9	92.7	70.4	80.5	(220.19, 11.01)
6	IV	102.9	69.9	95.5	71.1	80.6	(210.76, 10.54)
	1981						
7	I	98.9	70.9	96.6	73.7	81.5	(205.44, 10.27)
8	II	94.4	70.4	95.4	74.8	83.4	(219.75, 10.99)
9	III	94.8	68.5	93.0	75.1	84.1	(231.80, 11.59)
10	IV	98.1	68.8	93.4	75.5	83.8	(224.93, 11.25)
	1982						
11	I	96.5	69.0	92.3	76.3	84.9	(233.05, 11.65)
12	II	96.1	68.6	89.7	76.6	85.4	(244.14, 12.21)
13	III	93.9	68.2	86.8	76.2	85.1	(258.62, 12.93)
	1982						
* 14	IV	95.9	67.4	86.4	75.1	85.1	(260.22, 13.01)
	1983						
15	II	106.1	67.5	87.7	74.3	83.9	(235.67, 11.78)

x_1 = petroleum, x_2 = chemical, x_3 = iron and steel,
x_4 = electrical machinery, x_5 = transport machinery
 * = checking data.

and use (4.31) to check how well the inferred fuzzy values for samples 14 and 15 agree with the actual fuzzy output.

Since we are using fuzzy data, the following solutions to the min problem (4.23) and max problem (4.25) are obtained.

(1) Min

$$A_0^* = (626.1, 0)_L, \qquad A_1^* = (-1.50, 0)_L$$
$$A_2^* = (6.28, 0)_L, \qquad A_3^* = (-6.32, 0.21)_L \Bigg\} \qquad (4.32)$$
$$A_4^* = (-1.85, 0)_L, \qquad A_5^* = (0.45, 0)_L$$

Table 4.5. Least Upper and Greatest
Lower Bounds for Inferred Fuzzy Output

Sample	Min $\bar{Y}_i^* = (\bar{y}_i^*, \bar{e}_i^*)$	Max $\underline{Y}_i^* = (\underline{y}_i^*, \underline{e}_i^*)$
1	(209.0, 23.1)	(212.2, 2.6)
2	(243.5, 21.7)	(246.0, 2.4)
3	(250.3, 20.8)	(251.3, 2.2)
4	(226.6, 19.9)	(225.1, 2.2)
5	(226.8, 19.2)	(227.2, 2.2)
6	(212.3, 19.8)	(212.8, 2.2)
7	(213.2, 20.0)	(211.8, 2.3)
8	(223.3, 19.8)	(221.3, 2.3)
9	(225.6, 19.3)	(224.4, 2.4)
10	(219.2, 19.4)	(218.8, 2.4)
11	(228.8, 19.1)	(228.4, 2.4)
12	(243.0, 18.6)	(243.0, 2.4)
13	(262.7, 18.0)	(262.0, 2.4)

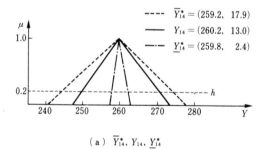

(a) $\bar{Y}_{14}^*$, Y_{14}, $\underline{Y}_{14}^*$

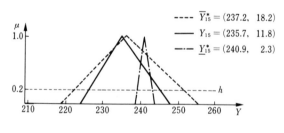

(b) $\bar{Y}_{15}^*$, Y_{15}, $\underline{Y}_{15}^*$

Fig. 4.4. Relation between Given Fuzzy Data Y_i and Greatest Lower ($\underline{Y}_i^*$) and Least
Upper ($\bar{Y}_i^*$) Bounds of Fuzzy Numbers for Checking Data (Samples 14 and 15)

(2) Max

$$A_0^* = (556.3, 0)_L, \qquad A_1^* = (-1.12, 0)_L$$
$$A_2^* = (6.34, 0)_L, \qquad A_3^* = (-6.41, 0)_L \qquad (4.33)$$
$$A_4^* = (-2.28, 0.03)_L, \qquad A_5^* = (1.28, 0)_L$$

Actually, $L(x) = 1 - |x|$ is assumed to be a triangular fuzzy number, and we let $h = 0.2$.

The inferred fuzzy outputs $\bar{Y}_i^*$ and $\underline{Y}_i^*$ of the linear possibility models from the min and max problems are shown in Table 4.5. From Tables 4.4 and 4.5, we get

$$\bar{Y}_i^* \underset{0.2}{\supseteq} Y_i \underset{0.2}{\supseteq} \underline{Y}_i^*, \qquad (4.34)$$

and $\bar{Y}_i^*$ and $\underline{Y}_i^*$ show the least upper and greatest lower bounds. The relationship between the inferred outputs for samples 14 and 15 and the given outputs is shown in Fig. 4.4. As can be seen from the figure, the solutions are such that Y_i is included within $\bar{Y}_i^*$, and Y_i includes $\underline{Y}_i^*$.

4.4 SUPPLEMENTARY NOTE

Methods for regression analysis using linear possibility systems were formulated in 1980,[3] and since then various types of regression analysis have been done.[4-8] The formulation is still being discussed and applications still being realized. However, the formulation remains new, so we can think of many points that must be discussed. When we think about the recently recognized need for bringing in advice from experts on data analysis, we see that fuzzy data is one form of expert knowledge. In addition, since regression analysis brings us back to LP problems, solutions are obtained by means of constraint on the coefficients that are introduced from the knowledge of experts. Finally, since fuzzy regression analysis provides interval inference, it is a robust form of analysis.

REFERENCES

(1) Dubois, D., and Prade, H., *Fuzzy Sets and Systems, Theory and Applications*, Academic Press, Cambridge, Mass., pp. 53–57 (1980).

(2) Zadeh, L.A., "The Concept of a Linguistic Variable and Its Application to Approximation Reasoning—I," *Information Sciences*, **8**, pp. 199–249 (1975).

(3) Tanaka, H., Uejima, S., and Asai, K., "Linear Regression Analysis with Fuzzy Model, "IEEE Transactions on Systems, Man and Cybernetics, SMC–12, **6**, pp. 903–907 (1982).

(4) Tanaka, H., Shimomura, T., Watada, J., and Asai, K., "Fuzzy Linear Regression Analysis of the Number of Staff in Local Government," FIP–84 at Kauai, Hawaii, July 22–26 (1984).

(5) Tanaka, H., "Fuzzy Data Analysis by Possibilistic Linear Models," *Fuzzy Sets and Systems*, **24**, pp. 363–375 (1987).

(6) Tanaka, H., and Watada, J., Possibilistic Linear Systems and Their Applications to the Linear Regression Models," *Fuzzy Sets and Systems*, **27**, pp. 275–289 (1988).

(7) Tanaka, H., Hayashi, I., and Watada, J., "Possibilistic Linear Regression Analysis for Fuzzy Data," *European Journal of Operational Research*, **40**, pp. 389–396 (1989).

(8) Tanaka, H., and Ishibuchi, H., "Identification of Possibilistic Linear Systems by Quadratic Membership Functions of Fuzzy Parameters," *The Proceedings of 3rd IFSA Congress*, pp. 516–519 (1989).

(9) Zheng, C. B., "On Applications of Fuzzy Linear Regression Models," working paper of Technical Institute of Ming Tsi, Taiwan (1982).

(10) Heshmaty, B., and Kandel, A., "Fuzzy Linear Regression and Its Applications to Forecasting in Uncertain Environment," *Fuzzy Sets and Systems*, **15**, pp. 159–191 (1985).

Chapter 5

STATISTICAL DECISION MAKING

When dealing with decision making problems created by probabilistic events, it is not uncommon to find the ambiguity of human subjectivity in the conditions. In these circumstances, fuzzy statistical decision making methods are formulated based on ideas resembling those of Bayes theory. Here we will make a comparative analysis of the value and amount of information for both probabilistic and fuzzy information; furthermore, we will discuss methods for using these kinds of information in discrimination problems under fuzzy conditions.

5.1 FUZZY PROBABILITY AND FUZZY ENTROPY

In ambiguous circumstances in which probabilistic events are the problem, cases in which the events themselves are actually ambiguous are not uncommon. An example is having to decide whether, if we think a product will sell well, we should advertise in such a way as to make the sales continue; or whether, if we think the product will not sell well, we should do research to find products that will sell better. Even if we know to a certain extent the probability distribution for the number of sales, the question remains as to what *sell well* and *not sell well* mean in concrete terms. If, for example, we say that making over 1,500 sales is selling well and making under 1,000 sales is selling poorly, whether making sales between 1,000 and 1,500 means *selling fairly well* or *selling somewhat poorly* will vary with the subjective judgment of the proprietor. This type of problem is actually surprisingly common. The concept of probability with fuzzy events is effective in these circumstances as a method for dealing with these conditions quantitatively.

First, let the standard probability space be (Ω, K, P). Here, Ω is the sample space, K the (complete) addition family from a subset of Ω and P the probability measure. The fuzzy set on Ω is called a *fuzzy event* here. Now let a ordinary event be $E \in K$. If we express the characteristic function as χ_E, the probability that event E will occur is expressed as follows:

$$P(E) = \int_\Omega \chi_E(\omega)\, dP; \qquad \chi_E(\omega) = \begin{cases} 1; & \omega \in E \\ 0; & \omega \notin E. \end{cases} \qquad (5.1)$$

Now let F be a fuzzy event, and if its membership function is μ_F, the probability that fuzzy event F will occur is defined as

$$P(F) = \int_\Omega \mu_F(\omega)\, dP; \qquad \mu_F(\omega): \Omega \to [0, 1], \qquad (5.2)$$

as an extension of (5.1).

If we think of the ordinary characteristic function as being extended to the membership function, we exchange the characteristic function in (5.1) for the membership function from the same point of view, and (5.2) can be interpreted as an extension of the probability of occurrence of the fuzzy event. This is the definition of the probability of a fuzzy event according to Zadeh.[1]

If we let the discrete sample space be $\Omega = \{\omega_1, \ldots, \omega_n\}$, the probability of fuzzy event F on Ω can be expressed by

$$P(F) = \sum_{i=1}^{n} \mu_F(\omega_i) P(\omega = \omega_i). \qquad (5.3)$$

Since this is a summation of the probability $P(\omega = \omega_i)$ multiplied by the degree to which ω_i belongs to fuzzy event F, it can be seen that it is a concept that is easy to accept intuitively as the probability that event F will occur.

If we say that there are two fuzzy events, A and B, the probability of the fuzzy events has the same properties as in the case of ordinary probability. That is, properties such as

(1) when $A \subset B$, $P(A) \leq P(B)$
(2) letting the complement of A be $\bar{A}$, $P(\bar{A}) = 1 - P(A)$
(3) $P(A \cup B) = P(A) + P(B) - P(A \cap B)$

arise. It can easily be confirmed that these properties arise based on operations with fuzzy sets defined by operations with membership functions.

Furthermore, the product of fuzzy events is defined as

$$A \cdot B \leftrightarrow \mu_{A \cdot B} = \mu_A \cdot \mu_B,$$

when A and B are independent, if and only if

$$P(A \cdot B) = P(A) \cdot P(B)$$

arises. In addition, if we let $P(A, B) \equiv P(A \cdot B)$, the conditional probability of a fuzzy event is defined as

$$P(A \mid B) = P(A, B)/P(B); \qquad P(B) > 0,$$

which is a natural extension of standard conditional probability.[2]

Next, let us discuss various types of entropy that are defined as measures for the degree of ambiguity of the circumstances in which a fuzzy event exists.[3] For simplicity, let $\Omega = \{\omega_1, \ldots, \omega_n\}$, and let a fuzzy event on Ω be A.

First, regardless of probability, the entropy for the ambiguity of the fuzzy set itself is defined as

$$\tilde{H}(A) = -k \sum_{i=1}^{n} \{\mu_A(\omega_i) \log \mu_A(\omega_i) + \mu_{\bar{A}}(\omega_i) \log \mu_{\bar{A}}(\omega_i)\}; \qquad k > 0. \quad (5.4)$$

In addition, the probabilistic entropy on a fuzzy set is defined as

$$H^P(A) = -\sum_{i=1}^{n} \mu_A(\omega_i) P(\omega = \omega_i) \log P(\omega = \omega_i). \quad (5.5)$$

Furthermore, the entropy for the occurrence of the fuzzy event itself is defined as

$$H(A) = -P(A) \log P(A) - P(\bar{A}) \log P(\bar{A}). \quad (5.6)$$

Example 5.1.   Let us consider the fuzzy event "a large roll" when dice are thrown. In this case $\Omega = \{1, \ldots, 6\}$ and $\mu_A(\omega_i)$ is given as in Table 5.1. Let the

Table 5.1.   Probability Distribution P and Membership Function μ_A

ω_i	1	2	3	4	5	6
$P(\omega = \omega_i)$	1/6	1/6	1/6	1/6	1/6	1/6
$\mu_A(\omega_i)$	0	0	0	0.2	0.6	1.0

logarithm be base e. The results of the calculations for (5.4)–(5.6) are

$$\tilde{H}(A) = -0.2\log 0.2 - 0.6\log 0.6 - 0.8\log 0.8 - 0.4\log 0.4$$

$$= 1.173; \qquad k = 1,$$

$$H^P(A) = -0.2 \cdot \frac{1}{6}\log\frac{1}{6} - 0.6 \cdot \frac{1}{6}\log\frac{1}{6} - \frac{1}{6}\log\frac{1}{6} = 0.538,$$

$$H(A) = -0.3\log 0.3 - 0.7\log 0.7 = 0.611.$$

Here, $0\log 0 = 0$.

The maximum values of the various entropies for all possible sets for A are

$$\max_A \tilde{H}(A) = 4.159, \qquad \max_A H^P(A) = 1.792, \qquad \max_A H(A) = 0.693,$$

and the percentage of the values of three types of entropies for the maximum values are as follows: 28.2% for $\tilde{H}(A)$, 30.0% for $H^P(A)$ and 88.2% for $H(A)$. In other words, the degree of ambiguity for the occurrence of the fuzzy events in this example is higher than in other cases. Of course, the entropy used will vary with the problem to be coped with, and what this example shows us is that even when identical circumstances are established, the degree of ambiguity naturally changes and must be interpreted if the point of view changes.

In the above we explained the probability concept of a fuzzy event according to the Zadeh definition. Research introducing fuzzy random variables is also being carried out.[4-6]

5.2   FUZZY-BAYES DECISION MAKING

Taking a problem like the one at the beginning of the last section as an example, we will first explain the concepts of fuzzy statistical decision making. Let us say that there are two possible fuzzy conditions for the sales of a certain product, "large sales" (F_1) and "poor sales" (F_2). Also, there are two possible actions to be taken this year, "advertise" (A_1) and "research products" (A_2). It is also possible to view these actions as forms of fuzzy action: "put greater effort into advertising" and "put greater effort into researching products." Given these fuzzy conditions and (fuzzy) actions, we shall say that the approximate utility values given are those in Table 5.2. Now we can formulate a fuzzy statistical decision making problem for the problem we have roughly established from a global viewpoint.

Table 5.2. Fuzzy Utility Function $U(A_i, F_j)$

(Fuzzy) Action ╲ Fuzzy Condition	Large Sales (F_1)	Poor Sales (F_2)
Advertise A_1)	800	−300
Research (A_2)	500	200

Here we will follow a Bayes theory-like decision method, and we will consider the action, A_1 or A_2 from Table 5.2, that produces the highest expected utility to be the most appropriate. However, in order to calculate the expected utilities, we must be able to calculate the probability of occurrence of fuzzy conditions F_1 and F_2. The basic idea here is to express the probability of occurrence using the concept of the probability of fuzzy events. In order to make the theoretical structure easy to understand, all of the spaces in this section are taken to be discrete finite sets.

Example 5.2. Let's say that a problem like that in Table 5.2 is given. We divide next year's sales into numerical blocks of 100 to form the elements of the basic space Ξ for establishing the fuzzy conditions so that $\Xi = \{800, 900, \ldots, 1700\}$, and let the *a priori* probability be $\xi(\theta_i)$. We will say that the *a priori* distribution $\xi(\theta_i)$ and fuzzy conditions F_1 and F_2 on Ξ are given as in Table 5.3. In this case, we get the probability of occurrence from equation 5.3 as follows:

$$P(F_k) = \sum_{i=1}^{10} \mu_{F_k}(\theta_i)\xi(\theta_i) = 0.5; \qquad k = 1, 2.$$

Let's also say that information about this year's sales of another product, which will serve as a leading indicator for the sales of the product in the problem, is obtainable. If we say that a particular θ_i will be realized next year, we can let, for example, the figures in Table 5.4 be given for the conditional probability $f(x_j \mid \theta_i)$, which shows the probability for the appearance of x_j number of sales this year for the product that is the leading indicator, when the

Table 5.3. Membership Function μ_{F_i} on θ and *a priori* Distribution ξ

θ_i	θ_1	θ_2	θ_3	θ_4	θ_5	θ_6	θ_7	θ_8	θ_9	θ_{10}
$\mu_{F_1}(\theta_i)$	0	0	0	0.2	0.4	0.6	0.8	1	1	1
$\mu_{F_2}(\theta_i)$	1	1	1	0.8	0.6	0.4	0.2	0	0	0
$\xi(\theta_i)$	0.05	0.05	0.1	0.1	0.2	0.2	0.1	0.1	0.05	0.05

Table 5.4. Values for Conditional Probability $f(x_j \mid \theta_i)$

θ_i \ x_j	x_1	x_2	x_3	x_4	x_5	x_6	x_7	x_8	x_9	x_{10}	x_{11}	x_{12}	x_{13}	x_{14}
θ_1	0.1	0.2	0.4	0.2	0.1									
θ_2		0.1	0.2	0.4	0.2	0.1						0		
$\vdots$			0											
θ_{10}										0.1	0.2	0.4	0.2	0.1

Table 5.5. Membership Function μ_{M_j} for Fuzzy Event M_j

x_i	x_1	x_2	x_3	x_4	x_5	x_6	x_7	x_8	x_9	x_{10}	x_{11}	x_{12}	x_{13}	x_{14}
$\mu_{M_1}(x_i)$	0	0	0	0	0	0	0	0	0.2	0.4	0.6	0.8	1	1
$\mu_{M_2}(x_i)$	0	0	0.2	0.4	0.6	0.8	1	1	0.8	0.6	0.4	0.2	0	0
$\mu_{M_3}(x_i)$	1	1	0.8	0.6	0.4	0.2	0	0	0	0	0	0	0	0

observation space X is $X = \{x_j\} = \{500, 600, \ldots, 1800\}$. In this case we say that information about which x_j is correct is obtainable, or even if this type of correct information on sales volume is difficult to obtain, that information in the form expressed by the fuzzy sets "good sales" (M_1), "average sales" (M_2) and "poor sales" (M_3) is obtainable. Let these fuzzy sets for the fuzzy events on observation space X be given as in Table 5.5.

Before we solve this problem for Example 5.2, let us make a general for-formulation of the fuzzy-Bayes decision method we will use here. We will let our fuzzy decision making problems be expressed in terms of $\langle \mathscr{F}, \mathscr{A}, U, \xi, \Xi \rangle$. $\mathscr{F} = \{F_1, \ldots, F_r\}$ is the set of fuzzy conditions, and F_i is a fuzzy event on $\Xi = \{\theta_1, \ldots, \theta_n\}$. $\mathscr{A} = \{A_1, \ldots, A_1\}$ is the set of actions, and $U(.,.)$ is the utility function on $\mathscr{A} \times \mathscr{F}$. $\xi(\cdot)$ is the *a priori* probability distribution on Ξ. It is assumed that the partition satisfies the orthogonal condition for the sets of the fuzzy conditions; that is, we let $\sum_{i=1}^{r} \mu_{Fi}(\theta_k) = 1 (k = 1, \ldots, n)$ be given rise to. The fuzzy conditions in Example 5.2 can also be looked at in this way.

Using the fact that we can calculate the probability of occurrence for F_i using Equation (5.3), the most appropriate action A° is defined as

$$U(A^\circ) \equiv \max_i U(A_i), \qquad U(A_i) = \sum_j U(A_i, F_j)P(F_j). \tag{5.7}$$

In other words, we let the action that gives the maximum expected utility be the most appropriate action. Of course

$$\sum_i P(F_i) = 1$$

is satisfied by the orthogonal condition.

Next we will think about cases in which it is possible to use additional information in the above decision making problem. Let us say that conditional probability $f(x_j | \theta_i)$ for observation space $X = \{x_1, \ldots, x_m\}$ is given. When we know which x_j in X will arise, we will call X a probabilistic information source.

Let us say that information x_j has been obtained from probabilistic information source X. From Bayes's rules, we get

$$\xi(\theta_i | x_j) = f(x_j | \theta_i)\xi(\theta_i)/f(x_j) \tag{5.8}$$

for the posterior probability $\xi(\theta_i | x_j)$ for θ_i. Here, $f(x_j)$ is the marginal probability for x_j and is expressed as follows:

$$f(x_j) = \sum_i f(x_j | \theta_i)\xi(\theta_i).$$

As shown in Example 5.2, we can also consider the case in which information about a fuzzy event on observation space X is received, and in this case the fuzzy observation space is expressed as $\mathcal{M} = \{M_1, \ldots, M_q\}$. Here, the orthogonal condition

$$\sum_{j=1}^{q} \mu_{M_j}(x_k) = 1 \qquad (k = 1, \ldots, m)$$

is satisfied.

Even for fuzzy events, we can derive Bayes's rules similar to those in (5.8) from the definition of the probability of a fuzzy event. These are called fuzzy-Bayes rules, and turn out as follows.

$$P(F_k | x_j) = \sum_i \mu_{F_k}(\theta_i)f(x_j | \theta_i)\xi(\theta_i)/f(x_j), \tag{5.9}$$

$$P(\theta_k | M_j) = \sum_i \mu_{M_j}(x_i)f(x_i | \theta_k)\xi(\theta_k)/P(M_j), \tag{5.10}$$

$$P(F_k | M_j) = \sum_i \sum_p \mu_{F_k}(\theta_i)\mu_{M_j}(x_p)f(x_p | \theta_i)\xi(\theta_i)/P(M_j). \tag{5.11}$$

Here $P(M_j)$ is the marginal probability for M_j and is expressed as follows:

$$P(M_j) = \sum_i \mu_{M_j}(x_i)f(x_i).$$

In circumstances in which this kind of fuzzy-Bayes rule arises, $\mathcal{M}$ is called a fuzzy information source when we know which M_j will arise.

Table 5.6. Values for Fuzzy
Conditional Probability $P(F_k|x_j)$ and
Probability Distribution $f(x_j)$

x_j	$P(F_1|x_j)$	$P(F_2|x_j)$	$f(x_j)$
x_1	0	1	0.005
x_2	0	1	0.015
x_3	0	1	0.040
x_4	0.034	0.966	0.060
x_5	0.126	0.874	0.095
x_6	0.288	0.712	0.125
x_7	0.424	0.576	0.160
x_8	0.576	0.424	0.160
x_9	0.712	0.288	0.125
x_{10}	0.874	0.126	0.095
x_{11}	0.966	0.034	0.060
x_{12}	1	0	0.040
x_{13}	1	0	0.015
x_{14}	1	0	0.005

Table 5.7. Values for Fuzzy Conditional
Probability $P(F_k|M_j)$

Information M_j	M_1	M_2	M_3
$P(F_1|M_j)$	0.913	0.5	0.087
$P(F_2|M_j)$	0.087	0.5	0.913

From the data given in Example 5.2, we can calculate $P(F_k|x_j)$, $f(x_j)$ and $P(F_k|M_j)$. The results are shown in Tables 5.6 and 5.7. We then get the following values for $P(M_j)$:

$$P(M_1) = 0.151, \qquad P(M_2) = 0.698, \qquad P(M_3) = 0.151.$$

When it is possible to use probabilistic information source X and fuzzy information source $\mathcal{M}$, and when we have obtained the x_j and M_j information, optimal actions $A_{x_j}^\circ$ and $A_{M_j}^\circ$ are defined as

$$U(A_{x_j}^\circ|x_j) \equiv \max_i U(A_i|x_j),$$

$$U(A_i|x_j) = \sum_k U(A_i, F_k)P(F_k|x_j), \tag{5.12}$$

$$U(A_{M_j}^\circ|M_j) \equiv \max_i U(A_i|M_j),$$

$$U(A_i|M_j) = \sum_k U(A_i, F_k)P(F_k|M_j). \tag{5.13}$$

In other words, when we have obtained the information, we let the maximal expected utility represent the optimal action. However, if we try to estimate the value of the information before obtaining it, we are in the situation of not knowing what information can be obtained. This is because we know only that we can use the information sources.

Here we can use the marginal probability to get an expected value for an *a priori* estimate for the information source. That is, we let the *a priori* estimates for information sources X and $\mathcal{M}$ be

$$U(A_X^\circ) = \sum_j U(A_{x_j}^\circ \mid x_j) f(x_j) \tag{5.14}$$

and

$$U(A_{\mathcal{M}}^\circ) = \sum_j U(A_{M_j}^\circ \mid M_j) P(M_j), \tag{5.15}$$

respectively. And we define values of the probabilistic and fuzzy information sources as the increasing part of the expected utilities, which are anticipated by means of the information obtained:

$$V(X) = U(A_X^\circ) - U(A^\circ), \tag{5.16}$$

$$V(\mathcal{M}) = U(A_{\mathcal{M}}^\circ) - U(A^\circ). \tag{5.17}$$

A conceptional schematic diagram is given in Fig. 5.1.

If we do the calculations for Example 5.2, $A^\circ = A_2$, and we get

$$U(A^\circ) = 350, \qquad V(X) = 62.0, \qquad V(\mathcal{M}) = 34.8.$$

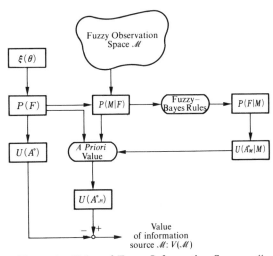

Fig. 5.1. Value of Fuzzy Information Source $\mathcal{M}$

In this case, the percentage of the existing value of information source $\mathcal{M}$ for information source X is about 56%. In other words, if we have the fuzzy information, we can think that the value of the information exists to some extent. Therefore, thinking from the standpoint of costs, there are cases when it is effective to use actual fuzzy information.

In the numerical example, the value of the probabilistic information source is greater than that of the fuzzy information source. We actually get

$$U(A^\circ_\mathcal{M}) = \sum_p \left\{ \max_i \sum_j U(A_i, F_j) \left[\sum_k P(F_j | x_k) \mu_{Mp}(x_k) f(x_k) \right] \right\}$$

$$\leq \sum_p \left\{ \sum_k \left[\max_i \sum_j U(A_i, F_j) P(F_j | x_k) \right] \mu_{M_p}(x_k) f(x_k) \right\}$$

$$= \sum_k U(A^\circ_{x_h} | x_k) f(x_k) = U(A^\circ_X),$$

which generally gives rise to a $V(X) \geq V(\mathcal{M})$ relationship. In addition, if we express the complete probabilistic information source (which gives the truth condition of Ξ without error) with X_∞ and the complete fuzzy information source (which tells us which fuzzy condition will arise with a probability of 1) with $\mathcal{M}_\infty$, the relationship

$$V(\mathcal{M}_\infty) \geq V(X_\infty) \geq V(X) \geq V(\mathcal{M}) \geq 0 \tag{5.18}$$

generally arises.[7]

We will now touch on the idea of the amount of fuzzy information. Using Equation (5.6), when x_j and M_j are given, the entropy for fuzzy event F_k is expressed by

$$H(F_k | x_j) = - P(F_k | x_j) \log P(F_k | x_j) - P(\bar{F}_k | x_j) \log P(\bar{F}_k | x_j),$$

$$H(F_k | M_j) = - P(F_k | M_j) \log P(F_k | M_j) - P(\bar{F}_k | M_j) \log P(\bar{F}_k | M_j).$$

We can then consider the a priori estimates as follows:

$$H(F_k | X) = \sum_j H(F_k | x_j) f(x_j) \tag{5.19}$$

and

$$H(F_k | \mathcal{M}) = \sum_j H(F_k | M_j) P(M_j), \tag{5.20}$$

and this is called the *conditional fuzzy entropy* under the condition in which the information sources can be used.

The entropies for fuzzy space $\mathcal{F}$ corresponding to the presence or absence of information are defined as

$$H(\mathcal{F}) = \tfrac{1}{2}\sum_k H(F_k), \qquad H(\mathcal{F}\,|\,X) = \tfrac{1}{2}\sum_k H(F_k\,|\,X),$$
$$H(\mathcal{F}\,|\,\mathcal{M}) = \tfrac{1}{2}\sum_k H(F_k\,|\,\mathcal{M}) \qquad\qquad\qquad (5.21)$$

The amounts of information for information sources X and $\mathcal{M}$ can then be defined as

$$I(X) = H(\mathcal{F}) - H(\mathcal{F}\,|\,X) \qquad\qquad (5.22)$$

and

$$I(\mathcal{M}) = H(\mathcal{F}) - H(\mathcal{F}\,|\,\mathcal{M}). \qquad\qquad (5.23)$$

In other words, we let the descending part of the expected entropy be the amount of information when we use the information source.

For Example 5.2, we get $H(\mathcal{F}) = 0.693$, $I(X) = 0.235$, and $I(\mathcal{M}) = 0.120$, and the percentage of the existing values of the amount of fuzzy information for the probabilistic information source is about 51%.

In general, $I(X) \geq 0$ and $I(\mathcal{M}) \geq 0$ arise, and properties such as

$$H(\mathcal{F}\,|\,X, Y) \leq H(\mathcal{F}\,|\,X) + H(\mathcal{F}\,|\,Y),$$
$$H(\mathcal{F}\,|\,\mathcal{M}_X, \mathcal{M}_Y) \leq H(\mathcal{F}\,|\,\mathcal{M}_X) + H(\mathcal{F}\,|\,\mathcal{M}_Y)$$

arise for two probabilistic information sources X and Y. Here $\mathcal{M}_x$ and $\mathcal{M}_y$ are fuzzy information sources on X and Y respectively.

In the above we have given a simple explanation of one part of the basic structure of fuzzy decision making methods composed along Bayes's lines. For the detailed theory, refer to references 3 and 7. There are various studies about fuzzy information done from the Bayes theory point of view taken in this section.[8-10]

5.3 FUZZY DISCRIMINATION METHODS

In this section we will also give an explanation using product sales as an example. As in Example 5.2 in the preceding section, there are two fuzzy conditions, the product either "sells well" or "does not sell well" and we say that there is information about the sales volume of another product that will serve as a leading index that we can use. Based on this information, we will think about which fuzzy condition we should more reasonably consider as likely to occur from the standpoint of the average error probability of discrimination. In other words, we are making the decision that minimalizes the average error discrimination probability, and this can be seen as a special form of fuzzy-Bayes decision making.

Let the *a priori* distribution $\xi(\theta_i)$ be defined on the space $\Xi = \{\theta_1, \ldots, \theta_n\}$. In order to simplify the discussion, we will let the fuzzy space be $\mathscr{F} = \{F_1, F_2\}$; and we let $\mathscr{F}$ satisfy the orthogonal conditions. Observation space X is generally thought of as being a continuous real-number space, and we let conditional probability density function $f(x \mid \theta_i)$ $(i = 1, \ldots, n)$ be given.

When probabilistic information x has been obtained, the probabilistic decision function $\delta_X(x)$, which shows us whether to make the judgment "condition F_1 will occur" or the judgment "condition F_2 will occur," is defined as

$$\delta_X(x) = (\delta_1(x), \delta_2(x)), \quad \delta_k(x): X \to [0, 1]; \qquad k = 1, 2. \tag{5.24}$$

Here, $\delta_k(x)$ shows the judgment that F_k will occur with probability $\delta_k(x)$, and it satisfies the relationship $\delta_1(x) + \delta_2(x) = 1$.

The probability of an error discrimination such as making judgment F_2 when F_1 occurs or making judgment F_1 when F_2 occurs can be given by using *a priori* probabilities $P(F_1)$ and $P(F_2)$, according to the Bayes system, and is expressed by

$$P_e(\delta_X) = P(F_1) \int_X \delta_2(x) f(x \mid F_1) \, dx + P(F_2) \int_X \delta_1(x) f(x \mid F_2) \, dx \tag{5.25}$$

This is called the *average error probability of discrimination*. Here $f(x \mid F_2)$ is derived as

$$f(x \mid F_k) = \sum_i \mu_{F_k}(\theta_i) f(x \mid \theta_i) \xi(\theta_i) / P(F_k); \qquad k = 1, 2$$

from the Zadeh definition. We then get

$$\delta_1^*(x) = \begin{cases} 1; & \Delta(x) \geq 0 \\ 0; & \Delta(x) < 0 \end{cases} \qquad \delta_2^*(x) = \begin{cases} 1; & \Delta(x) < 0 \\ 0; & \Delta(x) \geq 0 \end{cases} \tag{5.26}$$

for optimal discrimination rule $\delta_X^*(x) = (\delta_1^*(x), \delta_2^*(x))$ for $\Delta(x) = f(x \mid F_1) P(F_1) - f(x \mid F_2) P(F_2)$. In addition, we get

$$\delta_1^*(M_j) = \begin{cases} 1; & \Delta(M_j) \geq 0 \\ 0; & \Delta(M_j) < 0 \end{cases} \qquad \delta_2^*(M_j) = \begin{cases} 1; & \Delta(M_j) < 0 \\ 0; & \Delta(M_j) \geq 0 \end{cases} \tag{5.27}$$

for optimal discrimination rule $\delta_{\mathscr{M}}^*(M_j) = (\delta_1^*(M_j), \delta_2^*(M_j))$, when we obtain fuzzy information from fuzzy observation space $\mathscr{M} = \{M_1, \ldots, M_q\}$, which satisfies the orthogonal conditions defined on X.[11] The above discrimination rules are called *fuzzy discrimination rules*.

For Example 5.2, $f(x \mid F_k)$ and $P(M_j \mid F_k)$ turn out as shown in Tables 5.8 and 5.9, and the discrimination results, calculated on the basis of Equations (5.26) and (5.27), come out as in Tables 5.10 and 5.11. If we then calculate the average error discrimination probability when we set out to get probabilistic and fuzzy information, we come up with the following:

$$P_e(\delta_X^*) = P(F_1) \cdot \sum_{j=1}^{7} f(x_j \mid F_1) + P(F_2) \cdot \sum_{j=8}^{14} f(x_j \mid F_2) = 0.236,$$

$$P_e(\delta_{\mathcal{M}}^*) = P(F_1) \cdot P(M_3 \mid F_1) + P(F_2) \cdot \sum_{j=1}^{2} P(M_j \mid F_2) = 0.375.$$

In this case the error discrimination probability derived with fuzzy data is

Table 5.8. Values for Fuzzy Conditional Probability $f(x_j \mid F_k)$

Information x_j	$f(x_j \mid F_1)$	$f(x_j \mid F_2)$
x_1	0	0.010
x_2	0	0.030
x_3	0	0.080
x_4	0.004	0.116
x_5	0.024	0.166
x_6	0.072	0.178
x_7	0.136	0.184
x_8	0.184	0.136
x_9	0.178	0.072
x_{10}	0.166	0.024
x_{11}	0.116	0.004
x_{12}	0.080	0
x_{13}	0.030	0
x_{14}	0.010	0

Table 5.9. Values for Fuzzy Conditional Probability $P(M_j \mid F_k)$

Information M_j	$P(M_j \mid F_1)$	$P(M_j \mid F_2)$
M_1	0.276	0.026
M_2	0.698	0.698
M_3	0.026	0.276

Table 5.10. Discrimination Results with Probabilistic Information

Information x_j	$x_1 \sim x_7$	$x_8 \sim x_{14}$
Result	F_2	F_1

Table 5.11. Discrimination Results with Fuzzy Information

Information M_j	M_1, M_2	M_3
Result	F_1	F_2

larger than that derived using probabilistic data, but the ratio is about 1.59, so the fuzzy information does not compare unfavorably with the probabilistic information. In other words, even if the information for a discrimination problem is fuzzy information, it has a certain degree of value as information. From the point of view of costs, there are cases in which fuzzy information is more effective.

Finally, we will give a simple outline of the relationship between average error discrimination probability and number of observations. Let there be an independent probabilistic observation series $x(h) = (x_1, \ldots, x_h)$. That is,

$$f(x_1, \ldots, x_h \mid \theta_i) = \prod_{j=1}^{h} f(x_j \mid \theta_i); \qquad i = 1, \ldots, n$$

arises. When the entropy for the fuzzy state space while obtaining $x(h)$ is defined by

$$H_2(\mathscr{F}(x(h))) = -\sum_k P(F_k \mid x(h)) \log_2 P(F_k \mid x(h)), \qquad (5.28)$$

the relationship

$$P_e(\delta_X^*(x(h))) \le E\{H_2(\mathscr{F}(x(h)))\} \qquad (5.29)$$

arises. E is the symbol of expectation. Furthermore, for

$$\lambda = \int_X [f(x \mid F_1)f(x \mid F_2)]^{1/2} \, dx, \qquad (5.30)$$

$$q_k(h) = \sum_i \left(\frac{P(F_k)}{\xi(\theta_i)\mu_{F_k}(\theta_i)} \right)^{h-1}; \qquad k = 1, 2, \qquad (5.31)$$

the relationship

$$E\{H_2(\mathscr{F}(x(h)))\} \le 2C[P(F_1)P(F_2)]^{1/2} \cdot \lambda^h \sqrt{q_1(h)q_2(h)} \qquad (5.32)$$

arises.[11] Here, C is a constant, and the summation part of Equation (5.31) is the sum for the part for which $\xi(\theta_i)\mu_{Fk}(\theta_i)$ is positive. From the Schwartz inequality, if $f(x \mid F_i) = f(x \mid F_2)$ does not arise almost everywhere, it follows that $\lambda < 1$ and $\lim_{h \to \infty} \lambda_h = 0$. However, there is generally no guarantee that

$\lambda^h\sqrt{q_1(h)q_2(h)}$ will converge to zero. Therefore, even if there are an infinite number of observations, there is no guarantee that the average error discrimination probability will be zero. But the right side of Equation (5.32) is function of h, and Equation (5.32) should be helpful for judging to what extent information gathering is effective.

In addition, when the discrimination of whether state θ_1 or θ_2 of $\Xi = \{\theta_1, \theta_2\}$ will occur is performed using fuzzy information, h number of independent observations is expressed as $M(h) = (M_1, \ldots, M_h)$ and we get the following relationship:

$$P_e(\delta^*_{\mathcal{M}}(M(h))) \leq E\{H_2(\theta(M(h)))\} \leq 2C\lambda^h[\xi(\theta_1)\xi(\theta_2)]^{1/2}. \qquad (5.33)$$

If for all i we do not get $P(M_i|\theta_1) = P(M_i|\theta_2)$, we get

$$\lambda = \sum_{i=1}^{q} \{P(M_i|\theta_1)P(M_i|\theta_2)\}^{1/2} < 1,$$

and we can see that the average error discrimination probability converges to zero for an infinite number of observations.[12]

When both the states and the information are fuzzy, it is clear that a discussion similar to that for ordinary statistical discrimination theory will arise if we bring together the fuzzy-Bayes rules from the previous section and the problem structure from this section.

If we do the calculations for the upper bound of the average error discrimination probability for Example 5.2 for a single observation, we get

$$E\{H_2(\mathcal{F}(x))\} = 0.660, \qquad E\{H_2(\mathcal{F}(M))\} = 0.827.$$

The ratio of the upper bound for fuzzy information to that for probabilistic information is about 1.25. Other estimations for the upper bound can be made using an inequality that differs from that of (5.29). The discussion as to the level at which information gathering should be stopped is still going on. For a detailed discussion of the relationship between number of observations and average error discrimination probability see references (11) and (12).

REFERENCES

(1) Zadeh, L. A., "Probability Measures of Fuzzy Events," *Journal of Mathematical Analysis and Applications*, **22**, pp. 421–427 (1968).

(2) Nishida, T., and Takeda, E., *Fuzzy Sets and Their Applications*, pp. 21–28, Tokyo, Morikita Shuppan (1978) (in Japanese).

(3) Asai, K., and Negoita, C. V., eds., *Introduction to Fuzzy Systems Theory*, Tokyo, Ohmsha (1978) (in Japanese).

(4) Kwakernaak, H., "Fuzzy Random Variables-I, -II," *Information Sciences*, **15**, pp. 1–29 (1978); **17**, pp. 253–278 (1979).

(5) Kandel, A., and Byatt, W. J., "Fuzzy Processes," *Fuzzy Sets and Systems*, **4**, pp. 117–152 (1980).

(6) Miyakoshi, M., and Shimbo, M., "A Strong Law of Large Numbers for Fuzzy Random Variables," *Fuzzy Sets and Systems*, **12**, pp. 133–142 (1984).

(7) Tanaka, H., Okuda, T., and Asai, K., "Fuzzy Information and Decision in Statistical Model," Gupta, M. M., Ragade, R. K., and R. R., eds., in *Advances in Fuzzy Sets Theory and Applications*, pp. 303–320, Amsterdam North-Holland (1979).

(8) Gil, M. A., Lopez, M. T., and Gil, P. G., "Quality of Information; Comparison between Information Systems: 1. Non-Fuzzy States; 2. Fuzzy States," *Fuzzy Sets and Systems*, **15**, pp. 65–78, 129–145 (1985).

(9) Gil, M. A., Corral, N., and Gil, P. G., "The Fuzzy Decision Problem: An Approach to the Point Estimation Problem with Fuzzy Information," *European Journal of Operational Research*, **22**, pp. 26–34 (1985).

(10) Pardo, L., Information Energy of a Fuzzy Event and a Partition of Fuzzy Events, *IEEE Transactions on Systems, Man, and Cybernetics*, SMC-15, pp. 139–144 (1985).

(11) Asai, K., Tanaka, H., and Okuda, T., "On Discrimination of Fuzzy States in Probability Space," *Kybernetes*, **6**, 185–192 (1977).

(12) Okuda, T., Tanaka, H., and Asai, K., "Discrimination Problem with Fuzzy States and Fuzzy Information," in Zimmermann, H. J., Zadeh, L. A., and Gaines, R. R., *Fuzzy Sets and Decision Analysis*, pp. 97–106, Amsterdam North-Holland (1984).

Chapter 6

FUZZY QUANTIFICATION THEORY

In general terms, quantification methods are means of grasping data like human judgments and evaluations, which are not normally given numerical expression, in terms of quantities and understanding them. Actually, the recognition, judgment and evaluation activities that humans carry out are commonly expressed in qualitative linguistic terms such as *heavy*, *extremely light*, or *fast*. It would be easier to compare qualitative judgments and to learn the evaluative structure underneath them if we could replace qualitative expressions with numerical expressions. In order to accomplish this, we use quantification methods that are a type of multivariate analysis.

In Japan, the quantification theory proposed by Chikio Hayashi[1] in the 1950s is well known as a method for the quantification of qualitative judgments and evaluations. This quantification theory consists of four methods, I, II, III, and IV[2].

This series of quantification methods makes use of the values $\{1, 0\}$, which indicate $\{yes, no\}$ judgments, and analysis can be carried out by simple methods such as solving eigenvalue problems. Computer analysis is therefore easy, and these quantification methods are used for analysis of real problems in numerous areas such as management science, marketing, psychology, sociology, engineering, and medicine.

This chapter will describe methods for handling qualitative data using fuzzy set theory. The fuzzy quantification theory handled here will be explained in terms of the concept of fuzzy events,[3] using values on [0, 1] that express qualitative judgments.

6.1 CHARACTERISTICS OF FUZZY QUANTIFICATION THEORY

In this section, we will explain the handling of fuzzy data and fuzzy events that provides the basis for fuzzy quantification theory. Since the sample sets are commonly called *groups* in multivariate analysis, we will call the fuzzy sets that form the samples here *fuzzy groups*.

First let us consider a survey concerning electric razors made by two home electronics manufacturers, Company A and Company B, in order to find out primary purchasing factors. Both companies specified the features, price, and appearance of their product and asked that the degree of will to purchase and the degree of consideration of features, price, and appearance be specified in values over [0, 1]. For example, respondent *a*'s will to purchase Company A's razor was 0.8, and his degrees of consideration of features, price, and appearance were 0.9, 0.2, and 0.7 respectively. In this case we have a fuzzy set for liking Company A's razor, and the membership value for respondent *a* toward this fuzzy set is 0.8.

Using this kind of information means that an analysis can be made of the detailed information obtained from the people who responded to the survey, and it is possible to base the analysis on reality. The events defined by this kind of fuzzy set are called fuzzy events, as was discussed in Chapter 5.[4]

In order to review that definition, we will rewrite probability $P(A)$ of the fuzzy event determined by fuzzy set A over nth dimensional interval R^n, which is defined by degree of probability P, using Equation (5.2):

$$P(A) = \int_{R^n} \mu_A(x)\,dP$$
$$= E(\mu_A). \tag{6.1}$$

Here, $E(\mu_A)$ is the expected value of membership function μ_A.

Example 6.1.   The discrete probability

$$P = \{p^{x_1}, p^{x_2}, p^{x_3}, p^{x_4}, p^{x_5}\}$$
$$= \{0.1, 0.2, 0.4, 0.2, 0.1\}$$

on $\{x_1, x_2, x_3, x_4, x_5\}$ is defined. In this instance the probability $P(F)$ for fuzzy event

$$F = 0.6/x_1 + 1.0/x_2 + 0.5/x_3 + 0.2/x_4$$

can be calculated as follows:

$$P(F) = 0.1 \times 0.6 + 0.2 \times 1.0 + 0.4 \times 0.5 + 0.2 \times 0.2 = 0.5.$$

Using equation (6.1), the fuzzy mean and fuzzy variance for variable x can be calculated as follows:

$$m_A = \frac{1}{P(A)} \left\{ \int_{R^n} x \mu_A(x) \, dP \right\} \tag{6.2}$$

$$\sigma_A^2 = \frac{1}{P(A)} \left\{ \int_{R^n} (x - m_A)^2 \mu_A(x) \, dP \right\}. \tag{6.3}$$

Also, when we know that fuzzy event A is occurring—in other words, when we consider it to be within range of the fuzzy event—the probability of fuzzy event B, $P_A(B)$ is

$$P_A(B) = \frac{1}{P(A)} \left\{ \int_{R^n} \mu_B(x) \mu_A(x) \, dP \right\} = E_A(\mu_B). \tag{6.4}$$

Example 6.2.   When variable x in example 6.1 is $x_1 = 6$, $x_2 = 2$, $x_3 = 5$, $x_4 = 1$, $x_5 = 2$, the fuzzy mean and fuzzy variance work out as follows:

$$m_A = \frac{1}{0.5} \{6 \times 0.1 \times 0.6 + 2 \times 0.2 \times 1.0 + 5 \times 0.4 \times 0.5 + 1 \times 0.2 \times 0.2\}.$$

$$= 3.6$$

$$\sigma_A^2 = 3.04$$

The following relationships arise concerning the fuzzy event.

$$\sigma_A^2 = E_A\{(x - m_A)^2\}$$
$$= E_A(x^2) - E_A^2(x).$$

Here we will define the statistics (sample mean, sample variance) for the given sample $(x_1, \ldots, x_n)$, when we are concerned with fuzzy event A. The size of the fuzzy set is expressed as follows, using the elements of the set:

$$N(A) = \sum_{\omega=1}^{n} \mu_A(x_\omega). \tag{6.5}$$

Applying this idea of the size of fuzzy set $N(A)$ to the sample, we can define the sample mean m_A and variance σ_A^2 as follows:

$$m_A = \frac{1}{N(A)} \left\{ \sum_{\omega=1}^{n} x_\omega \mu_A(x_\omega) \right\} \tag{6.6}$$

$$\sigma_A^2 = \frac{1}{N(A)} \left\{ \sum_{\omega=1}^{n} (x_\omega - m_A)^2 \mu_A(x_\omega) \right\}. \tag{6.7}$$

Using these definitions, we will explain variation between groups, variation within groups, and total variation for fuzzy groups.

Let sample $x_\omega (\omega = 1, \ldots, n)$ be given and the membership function of fuzzy group $A_i (i = 1, \ldots, K)$ be defined by $\mu_{A_i}(x_\omega)$. In this instance, the total mean m and the mean m_{A_i} using fuzzy group A_i are expressed by the following equations:

$$m = \frac{1}{N} \left\{ \sum_{i=1}^{K} \sum_{\omega=1}^{n} x_\omega \mu_{A_i}(x_\omega) \right\}$$

$$m_{A_i} = \frac{1}{N(A_i)} \left\{ \sum_{\omega=1}^{n} x_\omega \mu_{A_i}(x_\omega) \right\}$$

Here we have

$$N = \sum_{i=1}^{k} N(A_i).$$

The total variation T, variation between fuzzy groups B, and variation within a fuzzy group E are defined as

$$T = \sum_{\omega=1}^{n} \sum_{i=1}^{K} (x_\omega - m)^2 \mu_{A_i}(x_\omega), \tag{6.8}$$

$$B = \sum_{\omega=1}^{n} \sum_{i=1}^{K} (m_{A_i} - m)^2 \mu_{A_i}(x_\omega), \tag{6.9}$$

$$E = \sum_{\omega=1}^{n} \sum_{i=1}^{K} (x_\omega - m_{A_i})^2 \mu_{A_i}(x_\omega), \tag{6.10}$$

respectively. The following relationship arises:[6]

$$T = B + E. \qquad (6.11)$$

Proof.

$$T = \sum_{\omega=1}^{n} \sum_{i=1}^{K} (x_\omega - m)^2 \mu_{A_i}(x_\omega)$$

$$= \sum_{\omega=1}^{n} \sum_{i=1}^{K} (m_{A_i} - m)^2 \mu_{A_i}(x_\omega) + \sum_{\omega=1}^{n} \sum_{i=1}^{K} (x_\omega - m_{A_i})^2 \mu_{A_i}(x_\omega)$$

$$= B + E.$$

We can understand that even in the case of a fuzzy event the total variation, variation between fuzzy groups, and variation within fuzzy groups can be separated one from another. This relationship shows that ideas used prior to now, like multivariate analysis derived from relationships such as maximum variance ratio, can easily be extended to fuzzy events.

6.2 FUZZY QUANTIFICATION THEORY I

The object of Fuzzy Quantification Theory I (qualitative regression analysis) is to find the relationship between qualitative descriptive variables, which are given in values on $[0, 1]$, and numerical object variables in the fuzzy groups given in the samples. Let us consider a simple example.

Example 6.3. We want to know the effect on certain shops's everyday sales of being surrounded by a shopping center. In order to do this we let the degree to which each shop handles everyday goods form a fuzzy set, and we analyze the relationship between sales and environment by means of the degree to which the area around each shop is a shopping center (Table 6.1).

If we express the data in Example 6.3 in general terms, we get the data handled by Fuzzy Quantification Theory I, as shown in Table 6.2. Fuzzy group B expresses the fuzzy set of the sample. y_ω is the objective function of sample ω, and $\mu_i(\omega)$ is the degree of response to qualitative category $i(i = 1, \ldots, K)$, which is given in values on $[0, 1]$.

Fuzzy Quantification Theory I means determining a linear function of categories

$$y(\omega) = \sum_{i=1}^{K} a_i \mu_i(\omega) \qquad (6.12)$$

Table 6.1. Data on Sales of Everyday Goods

	Y	X	Fuzzy Group B
	-------	-----------------	--------------------------------------
Sample	Sales	Shopping Center	Degree of Handling Everyday Goods
1	500	1.0	0.1
2	490	0.6	0.1
3	400	0.4	0.1
4	350	0.6	1.0
5	300	1.0	1.0
6	280	0.8	1.0
7	270	0.4	1.0
8	250	0.6	1.0
9	230	1.0	1.0
10	210	0.0	1.0
11	200	0.8	1.0
12	200	0.4	1.0
13	100	0.6	0.1
14	50	0.4	0.1
15	50	0.0	0.1

Table 6.2. Data Handled by Fuzzy Quantification Theory I

No. ω	External Standard y	Category $A_1 \cdots A_i \cdots A_K$			Fuzzy Group B
1	y_1	$\mu_1(1)$	$\cdots \mu_i(1) \cdots$	$\mu_K(1)$	$\mu_B(1)$
2	y_2	$\mu_1(2)$	$\cdots \mu_i(2) \cdots$	$\mu_K(2)$	$\mu_B(2)$
$\vdots$	$\vdots$	$\vdots$	$\vdots$	$\vdots$	$\vdots$
ω	y_ω	$\mu_1(\omega)$	$\cdots \mu_i(\omega) \cdots$	$\mu_K(\omega)$	$\mu_B(\omega)$
$\vdots$	$\vdots$	$\vdots$	$\vdots$	$\vdots$	$\vdots$
n	y_n	$\mu_1(n)$	$\cdots \mu_i(n) \cdots$	$\mu_K(n)$	$\mu_B(n)$

that best expresses the structure of the data, in other words, such that the object variation given and its error variance are minimized. In order to make the expressions simpler, we introduce the following matrix symbology:

$$y' = [y_\omega]' = [y_1, y_2, \ldots, y_n]$$

$$G = \begin{bmatrix} \mu_B(1) & & 0 \\ & \ddots & \\ 0 & & \mu_B(n) \end{bmatrix}$$

$$X = [\mu_i(\omega)] = \begin{bmatrix} \mu_1(1) & \cdots & \mu_i(1) & \cdots & \mu_K(1) \\ \vdots & & \vdots & & \vdots \\ \mu_1(\omega) & \cdots & \mu_i(\omega) & \cdots & \mu_K((\omega) \\ \vdots & & \vdots & & \vdots \\ \mu_1(n) & \cdots & \mu_i(n) & \cdots & \mu_K(n) \end{bmatrix}$$

$$a' = [a_i]' = [a_1, a_2, \ldots, a_K].$$

Here "'" denotes transposition.

Using this, the error variance σ_B^2 for fuzzy group B is

$$\sigma_B^2 = \frac{1}{N(B)} (y - Xa)'G(y - Xa).$$

From

$$\frac{\partial \sigma_B^2}{\partial a} = -2X'Gy + 2X'GXa = 0,$$

the category weight a that minimizes the error variance is given by the following equation:

$$a = (X'GX)^{-1}X'Gy. \tag{6.13}$$

If we find the category weight for Example 6.3 using Equation (6.13), we come up with the following:

(1) When we do not consider the degree to which the shops handle everyday goods as fuzzy group B, the relationship between sales y and the degree to which the area around the shop is a shopping center is

$$y = 224.0 + 70.9x.$$

The surroundings produce little effect.

(2) On the other hand, when we consider the degree to which the shops handle everyday goods as fuzzy group B, the relationship between sales y and effect of the surroundings x is

$$y = 149.8 + 190x,$$

and sales depend to a large extent on the area around the shop's being a shopping center.

In this way we can clearly educe that everyday goods depend on residences, that is, on shopping centers, through a consideration of the degree to which everyday goods are handled using fuzzy group B.

In order to find out the effect of each category on object variable y when changes in the other categories are fixed, let us explain fuzzy partial correlation coefficients.

The following fuzzy means and fuzzy covariances are defined for categories i and $y(\omega)$:

$$r_{ij} = \frac{\sigma_{ij}}{\sqrt{\sigma_{ii}\sigma_{jj}}}$$

$$r_{iy} = \frac{\sigma_{iy}}{\sqrt{\sigma_{ii}\sigma_{yy}}}.$$

Here $X_i(\omega) = a_i\mu_i(\omega)$. Using these covariances, fuzzy correlation coefficients r_{ij} and r_{iy} are

$$\bar{X}_i = \frac{1}{N}\left\{\sum_{r=1}^{M}\sum_{\omega=1}^{n} X_i(\omega)\mu_{B_r}(\omega)\right\}$$

$$\bar{y} = \frac{1}{N}\left\{\sum_{r=1}^{M}\sum_{\omega=1}^{n} y(\omega)\mu_{B_r}(\omega)\right\}$$

$$\sigma_{ij} = \frac{1}{N}\sum_{r=1}^{M}\sum_{\omega=1}^{n}(X_i(\omega) - \bar{X}_i)(X_j(\omega) - \bar{X}_j)\mu_{B_r}(\omega)$$

$$\sigma_{iy} = \frac{1}{N}\sum_{r=1}^{M}\sum_{\omega=1}^{n}(X_i(\omega) - \bar{X}_i)(y(\omega) - \bar{y})\mu_{B_r}(\omega)$$

$$\sigma_{yy} = \frac{1}{N}\sum_{r=1}^{M}\sum_{\omega=1}^{n}(y(\omega) - \bar{y})^2\mu_{B_r}(\omega),$$

and we make these elements into matrix $\boldsymbol{R}$:

$$\boldsymbol{R} = \begin{bmatrix} 1 & r_{12} & \cdots & r_{1K} & r_{1y} \\ r_{21} & 1 & \cdots & r_{2K} & r_{2y} \\ \vdots & \vdots & & \vdots & \vdots \\ r_{K1} & r_{K2} & \cdots & 1 & r_{Ky} \\ r_{y1} & r_{y2} & \cdots & r & 1 \end{bmatrix}.$$

The inverse matrix $\boldsymbol{R}^{-1}$ is written

$$\boldsymbol{R}^{-1} = \begin{bmatrix} r^{11} & r^{12} & \cdots & r^{1K} & r^{1y} \\ r^{21} & r^{22} & \cdots & r^{2K} & r^{2y} \\ \vdots & \vdots & & \vdots & \vdots \\ r^{y1} & r^{y2} & \cdots & r^{yK} & r^{yy} \end{bmatrix},$$

and objective variable y and its partial correlation coefficient
$r_{iy:1,2,\ldots,i-1,i+1,\ldots,K}$ are defined as follows:

$$r_{iy:1,2,\ldots,i-1,i+1,\ldots,K} = \frac{-r^{iy}}{\sqrt{r^{ii} \cdot r^{yy}}}.$$

This partial correlation coefficient shows the effect of variable i on the objective variable when the other variables are fixed.

6.3   FUZZY QUANTIFICATION THEORY II

The object of Fuzzy Quantification Theory II (qualitative discrimination analysis) as proposed in Watada *et al.* is to express several fuzzy groups in terms of qualitative descriptive variables. These qualitative descriptive variables take the form of values (membership values) on $[0, 1]$.

Table 6.3 shows the data handled by Fuzzy Quantification Theory II. The difference from Fuzzy Quantification Theory I lies in the fact that external standards are given by fuzzy groups $B_1, \ldots, B_M$.

The object of Fuzzy Quantification Theory II is to express, as well as possible using the linear equation of category weight a_i of category A_1, the structure of the external standard fuzzy groups on the real number axis:

$$y(\omega) = \sum_{i=1}^{K} a_i \mu_i(\omega); \qquad \omega = 1, \ldots, n. \tag{6.14}$$

In other words, it is determining the a_i that gives the best separation of the external standard fuzzy groups on the real number axis.

The degree of separation of the fuzzy groups is defined as the fuzzy variance ratio η^2, which is the ratio of the total variation T and variation between fuzzy

Table 6.3.   Data Handled by Fuzzy Quantification Theory II

No. ω	Fuzzy External Standard $B_1 \cdots B_M$	Category $A_1 \cdots A_i \cdots A_K$
1	$\mu_{B_1}(1) \cdots \mu_{B_M}(1)$	$\mu_1(1) \cdots \mu_i(1) \cdots \mu_K(1)$
2	$\mu_{B_1}(2) \cdots \mu_{B_M}(2)$	$\mu_1(2) \cdots \mu_i(2) \cdots \mu_K(2)$
$\vdots$	$\vdots$	$\vdots \quad \vdots \quad \vdots$
ω	$\mu_{B_1}(\omega) \cdots \mu_{B_M}(\omega)$	$\mu_1(\omega) \cdots \mu_i(\omega) \cdots \mu_K(\omega)$
$\vdots$	$\vdots$	$\vdots \quad \vdots \quad \vdots$
n	$\mu_{B_1}(n) \cdots \mu_{B_M}(n)$	$\mu_1(n) \cdots \mu_i(n) \cdots \mu_K(n)$

groups B from Equations (6.8) and (6.9):

$$\eta^2 = \frac{B}{T}. \tag{6.15}$$

In the following we will determine a_i for Linear Equation (6.14), which maximizes fuzzy variance ratio η^2.

The fuzzy mean $\bar{y}_{B_r}$ within fuzzy group B_r for value $y(\omega)$ for the linear equation and total fuzzy mean y come out as follows:

$$\bar{y}_{B_r} = \frac{1}{N(B_r)} \left\{ \sum_{\omega=1}^{n} y(\omega) \mu_{B_r}(\omega) \right\}; \qquad r = 1, \dots, M.$$

$$\bar{y} = \frac{1}{N} \left\{ \sum_{r=1}^{M} \bar{y}_{B_r} N(B_r) \right\}$$

Fuzzy mean $\bar{\mu}_i^r$ within each fuzzy group B_r for the membership value of category A_i and total fuzzy mean $\bar{\mu}_i$ are expressed as follows:

$$\bar{\mu}_i^r = \frac{1}{N(B_r)} \left\{ \sum_{\omega=1}^{n} \mu_i(\omega) \mu_{B_r}(\omega) \right\}; \qquad i = 1, \dots, K, \qquad r = 1, \dots, M$$

$$\bar{\mu}_i = \frac{1}{N} \left\{ \sum_{r=1}^{M} \bar{\mu}_i^r N(B_r) \right\}; \qquad i = 1, \dots, K.$$

In order to simplify the writing, the (Mn, K) matrices A, $\bar{A}_G$, and $\bar{A}$ for $\mu_i(\omega)$, $\bar{\mu}_i^r$, and μ_i are defined as follows:

$$A = \begin{bmatrix} \mu_1(1) & \cdots & \mu_i(1) & \cdots & \mu_K(1) \\ \vdots & & \vdots & & \vdots \\ \mu_1(\omega) & \cdots & \mu_i(\omega) & \cdots & \mu_K(\omega) \\ \vdots & & \vdots & & \vdots \\ \mu_1(n) & \cdots & \mu_i(n) & \cdots & \mu_K(n) \\ \mu_1(1) & \cdots & \mu_i(1) & \cdots & \mu_K(1) \\ \vdots & & \vdots & & \vdots \\ \mu_1(n) & \cdots & \mu_i(n) & \cdots & \mu_K(n) \end{bmatrix};$$

$$\bar{A}_G = \begin{bmatrix} \bar{\mu}_1^1 & \cdots & \bar{\mu}_i^1 & \cdots & \bar{\mu}_K^1 \\ \vdots & & \vdots & & \vdots \\ \bar{\mu}_1^1 & \cdots & \bar{\mu}_i^1 & \cdots & \bar{\mu}_K^1 \\ \vdots & & \vdots & & \vdots \\ \bar{\mu}_1^1 & \cdots & \bar{\mu}_i^1 & \cdots & \bar{\mu}_K^1 \\ \bar{\mu}_1^2 & \cdots & \bar{\mu}_i^2 & \cdots & \bar{\mu}_K^2 \\ \vdots & & \vdots & & \vdots \\ \bar{\mu}_1^M & \cdots & \bar{\mu}_i^M & \cdots & \bar{\mu}_K^M \end{bmatrix};$$

$$\bar{A} = \begin{bmatrix} \bar{\mu}_1 & \cdots & \bar{\mu}_i & \cdots & \bar{\mu}_K \\ \vdots & & \vdots & & \vdots \\ \bar{\mu}_1 & \cdots & \bar{\mu}_i & \cdots & \bar{\mu}_K \\ \vdots & & \vdots & & \vdots \\ \bar{\mu}_1 & \cdots & \bar{\mu}_i & \cdots & \bar{\mu}_K \\ \bar{\mu}_1 & \cdots & \bar{\mu}_i & \cdots & \bar{\mu}_K \\ \vdots & & \vdots & & \vdots \\ \bar{\mu}_1 & \cdots & \bar{\mu}_i & \cdots & \bar{\mu}_K \end{bmatrix}.$$

In addition, the K dimensional row vector a for category weight a_i and the (Mn, Mn) diagonal matrix G formed from membership value μ_{B_r} are defined as follows:

$$a' = \begin{bmatrix} a_1 & \cdots & a_i & \cdots & a_K \end{bmatrix}$$

$$G = \begin{bmatrix} \mu_{B_1}(1) & & & & & \\ & \ddots & & & & 0 \\ & & \mu_{B_1}(n) & & & \\ & & & \mu_{B_2}(1) & & \\ & 0 & & & \ddots & \\ & & & & & \mu_{B_M}(n) \end{bmatrix}.$$

Using these, the total variation T and variation between fuzzy groups B from Equations (6.8) and (6.9) can be written as follows:

$$T = a'(A - \bar{A})'G(A - \bar{A})a \tag{6.16}$$

$$B = a'(\bar{A}_G - \bar{A})'G(\bar{A}_G - \bar{A})a. \tag{6.17}$$

We obtain the following relationship if we substitute Equations (6.16) and (6.17) for Equation (6.15) and partially differentiate by a:

$$\{G^{1/2}(\bar{A}_G - \bar{A})\}'\{G^{1/2}(\bar{A}_G - \bar{A})\} a$$
$$= \eta^2 \{G^{1/2}(A - \bar{A})\}'\{G^{1/2}(A - \bar{A})\} a. \tag{6.18}$$

If we now define S_G and S by the (K, K) matrix as

$$S_G = \{G^{1/2}(\bar{A}_G - \bar{A})\}'\{G^{1/2}(\bar{A}_G - \bar{A})\}$$

$$S = \{G^{1/2}(A - \bar{A})\}'\{G^{1/2}(A - \bar{A})\},$$

we can decompose S using triangular matrix Δ as $S = \Delta'\Delta$, and we get the following:

$$[(\Delta')^{-1}S_G \Delta^{-1}] \Delta a = \eta^2 \Delta a.$$

Table 6.4. Results of Survey on Purchasing Behavior in Buying Electric Razors

Respondent No.	External Standards		Items						
	Company A Product	Company B Product	X_1: Price		X_2: Appearance		X_3: Features		
			Consider	Don't Consider	Consider	Don't Consider	Check Thoroughly	Check	Don't Check
	B_1	B_2	A_{11}	A_{12}	A_{21}	A_{22}	A_{31}	A_{32}	A_{33}
①	0.5	0.5	0.8	0.2	0.9	0.0	0.0	0.8	0.1
②	0.8	0.2	0.1	0.7	0.8	0.2	1.0	0.0	0.0
③	1.0	0.9	0.9	0.0	0.0	1.0	0.8	0.1	0.0
④	0.7	0.2	0.0	1.0	0.0	1.0	1.0	0.0	0.0
⑤	0.2	0.7	1.0	0.0	0.6	0.4	0.0	0.3	0.5
⑥	0.0	1.0	0.0	0.7	0.0	0.7	0.0	0.0	1.0
⑦	0.2	0.8	0.0	1.0	0.0	1.0	0.0	1.0	0.0
⑧	0.0	0.9	0.2	0.7	0.7	0.3	0.0	0.2	0.6
⑨	0.3	0.7	0.0	1.0	1.0	0.0	0.0	1.0	0.0
⑩	0.0	1.0	0.0	1.0	1.0	0.0	0.0	0.0	1.0

Note: correspondence with text: $n = 10$, $K = 2 + 2 + 3 = 7$, $M = 2$

112

Because of this, category a for Equation (6.14), which maximizes fuzzy variance ratio η^2, can be obtained from eigenvector Δa, which maximizes eigenvalue η^2 of the matrix $[(\Delta')^{-1} S_G \Delta^{-1}]$.

Example 6.4. Home electronics producers Company A and Company B surveyed 10 students about their desire to buy one of the two companies' similar electric razors, based on features, price, and appearance. The data obtained are shown in Table 6.4. Using the degree of desire to buy each product as the external standard, let us analyze this problem using Fuzzy Quantification Theory II.

Tables 6.5 and 6.6 give the category weights and partial correlation coefficients that were obtained as results of the analysis. Figure 6.1 expresses in graphic form the membership values of the external standards and of each sample. As can be seen from Fig. 6.1, the structure of levels of those who like Company A's product versus those who like Company B's is shown very well.

In addition, the preference of consumers for Company A or Company B can be anticipated from these results. For example, let Table 6.7 be one consumer's

Table 6.5. Category Weights in the Survey about Electric Razors

Item		Category Weight	
price	consider	a_{11}	0.43
	don't consider	a_{12}	0.07
appearance	consider	a_{21}	−0.24
	don't consider	a_{22}	−0.28
features	check thoroughly	a_{31}	1.41
	check	a_{32}	0.89
	don't check	a_{33}	0.62

Table 6.6. Obtained Partial Correlation Coefficients for the Electric Razor Survey

Item	External Standard
price	0.417
appearance	0.072
features	0.650

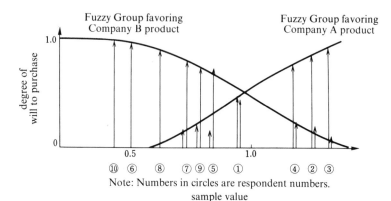

Note: Numbers in circles are respondent numbers.
sample value

Fig. 6.1. Samples and Membership Values for the Electric Razor Survey

Table 6.7. Data for New Respondents to the
Electric Razor Survey

	Item	Category Weight
price	consider	1.0
	don't consider	0.0
appearance	consider	0.0
	don't consider	1.0
features	check thoroughly	0.0
	check	1.0
	don't check	0.0

answers to the survey. From Equation (6.14) this consumer's sample value is
0.77, and his degrees of desire to buy Company A's and Company B's products
are about 0.7 and 0.3, respectively, as obtained from the graph in Fig. 6.1.

6.4 FUZZY QUANTIFICATION THEORY III

Fuzzy Quantification Theory III is a method in which pattern classification is
done; similar methods were developed independently in various countries.
They are known by different names such as *dual scaling, correspondence
analysis, pattern classification*, etc., and are described in detail in a paper by
Tenenhaus and Young[8] and in a book by Lebart, *et al.*[9]

Fuzzy Quantification Theory III is a method in which, if samples are taken
from *young* people, we think of these samples as being elements of fuzzy set *B*,

"*young* people," and attempt quantitatively to classify each sample ω and category by considering the membership values of this fuzzy set. In this case, the response to each category is given a degree of attribution, not on $\{0,1\}$ but on $[0,1]$, and we determine whether the response pattern for each sample differs. Especially when the response to the categories are given by $\{0,1\}$, we can think of the frequency of the patterns as the degree of attribution.

Example 6.5. Let us take a look at a simple example. Table 6.8 is the data collected from young people about shops and brands when they buy products. Pattern classification was done in order to learn about the purchasing behavior of young people. At this time, it was possible to grasp the consumer activities of young people by considering the degree of youth of each respondent as shown in Table 6.9.

Table 6.8. Data about Young People's Purchasing Activities

	Place of Purchase			Product Knowledge		Brand-Name Products		
Sample	Supermarkets	Specialty Shops	Department Stores	None	Much	Doesn't Matter	Usually Buy	Youthfulness
1	1			1		1		0.2
2		1			1		1	1.0
3			1	1			1	0.8
4			1	1			1	0.1
5		1			1	1		1.0
6			1		1		1	0.8
7			1	1		1		0.8

Table 6.9. Data Handled by Fuzzy Quantification Theory III

$$u_1 \cdots u_i \cdots u_K$$

	No. ω	Fuzzy Group B	Category $1 \cdots i \cdots K$	Total
v_1	1	$\mu_B(1)$	$\mu_1(1) \cdots \mu_i(1) \cdots \mu_K(1)$	m_1
v_2	2	$\mu_B(2)$	$\mu_1(2) \cdots \mu_i(2) \cdots \mu_K(2)$	m_2
$\vdots$	$\vdots$	$\vdots$	$\vdots \quad\quad \vdots \quad\quad \vdots$	$\vdots$
v_ω	ω	$\mu_B(\omega)$	$\mu_1(\omega) \cdots \mu_i(\omega) \cdots \mu_K(\omega)$	m_ω
$\vdots$	$\vdots$	$\vdots$	$\vdots \quad\quad \vdots \quad\quad \vdots$	$\vdots$
v_n	n	$\mu_B(n)$	$\mu_1(n) \cdots \mu_i(n) \cdots \mu_K(n)$	m_n

$$m_\omega = \Sigma\mu_i(\omega); \quad \omega = 1,\ldots,n$$

Table 6.9 shows a generalization of the data handled by Fuzzy Quantification Theory III. The object of Fuzzy Quantification Theory III is to give a mutually close numerical value to the samples resembling reactions, when real numerical values v_ω and u_i are assigned to samples ω and to the various categories i, and at the same time to give a close numerical value to resembling categories. In order to do this, the correlation coefficient of the category is used as an indicator.

The size of reactions for all categories for sample ω is defined as

$$m_\omega = \sum_{i=1}^{K} \mu_i(\omega).$$

Samples with a large membership value to fuzzy set B are evaluated highly in this analysis, but those with low ones are not given much consideration. As a result, the reaction for all of the data is defined as the product of the membership values to fuzzy set B and the above equation:

$$T = \sum_{\omega=1}^{n} m_\omega \mu_B(\omega).$$

Means u and v, variances σ_u^2 and σ_v^2, and covariance σ_{uv} for numerical values u_i and v_ω, based on fuzzy event B, come out as follows:

$$\bar{u} = \frac{1}{T} \left\{ \sum_{\omega=1}^{n} \sum_{i=1}^{K} \mu_i(\omega)\mu_B(\omega)u_i \right\}$$

$$\bar{v} = \frac{1}{T} \left\{ \sum_{\omega=1}^{n} m_\omega \mu_B(\omega)v_\omega \right\}$$

$$\sigma_u^2 = \frac{1}{T} \left\{ \sum_{\omega=1}^{n} \sum_{i=1}^{K} \mu_i(\omega)\mu_B(\omega)u_i^2 \right\} - \bar{u}^2$$

$$\sigma_v^2 = \frac{1}{T} \left\{ \sum_{\omega=1}^{n} m_\omega \mu_B(\omega)v_\omega^2 \right\} - \bar{v}^2$$

$$\sigma_{uv} = \frac{1}{T} \left\{ \sum_{\omega=1}^{n} \sum_{i=1}^{K} \mu_i(\omega)\mu_B(\omega)u_i v_\omega \right\} - \bar{u}\bar{v}.$$

Based on the conditions that $\bar{u} = 0$, and $\bar{v} = 0$, the problem here is to determine the u_i and v_ω that maximize correlation coefficient

$$\rho = \frac{\sigma_{uv}}{\sqrt{\sigma_u^2 \sigma_v^2}}.$$

A partial differentiation of correlation coefficients u_k and v_τ gives us

$$\frac{\partial \rho}{\partial u_k} = 0; \qquad k = 1, \ldots, K$$

$$\frac{\partial \rho}{\partial v_\tau} = 0; \qquad \tau = 1, \ldots, n.$$

If we do these calculations, we get the following:

$$\sum_{\omega=1}^{n} \mu_k(\omega)\mu_B(\omega)v_\omega = \rho \frac{\sigma_v}{\sigma_u} \sum_{\omega=1}^{n} \mu_k(\omega)\mu_B(\omega)u_k$$

$$\sum_{i=1}^{K} \mu_i(\tau)\mu_B(\tau)u_i = \rho \frac{\sigma_u}{\sigma_v} m_\tau \mu_B(\tau)v_\tau.$$

Eliminating v_τ, we get

$$\sum_{\omega=1}^{n} \sum_{i=1}^{K} \frac{\mu_B(\omega)}{m_\omega} \mu_k(\omega)\mu_i(\omega)u_i = \rho^2 \cdot \sum_{\tau=1}^{n} \mu_k(\tau)\mu_B(\tau)u_k; \qquad k = 1, \ldots, K. \quad (6.19)$$

In order to consolidate the above, we introduce the following notation:

$$b_k = \sum_{\omega=1}^{n} \mu_k(\omega)\mu_B(\omega); \qquad k = 1, \ldots, K$$

$$z_k = \sqrt{b_k u_k}; \qquad k = 1, \ldots, K$$

$$c_{ki} = \frac{1}{\sqrt{b_k b_i}} \sum_{\omega=1}^{n} \frac{\mu_B(\omega)}{m_\omega} \mu_k(\omega)\mu_i(\omega); \qquad k, i = 1, \ldots, K$$

$$c = [c_{ki}]$$

$$z' = [z_1, \ldots, z_k].$$

Using this, the matrix expression of equation (6.19) results in the following:

$$b_k = \sum_{\omega=1}^{n} \mu_k(\omega)\mu_B(\omega); \qquad k = 1, \ldots, K$$

$$z_k = \sqrt{b_k u_k}; \qquad k = 1, \ldots, K$$

$$c_{ki} = \frac{1}{\sqrt{b_k b_i}} \sum_{\omega=1}^{n} \frac{\mu_B(\omega)}{m_\omega} \mu_k(\omega)\mu_i(\omega); \qquad k, i = 1, \ldots, K$$

$$c = [c_{ki}]$$

$$z' = [z_1, \ldots, z_k]$$

$$cz = \rho^2 z$$

In other words, under the conditions $u = 0$ and $v = 0$, all we have to do is to find the maximum values for the eigenvalue equation. In addition, if we solve it based on the condition that v_r is $(1/\rho \cdot \sigma_u/\sigma_v) = 1$, we get

$$v_\tau = \frac{1}{m_\tau}\left\{\sum_{i=1}^{K} \mu_i(\tau)u_i\right\}.$$

We analyzed Example 6.5 by Fuzzy Quantification Theory III as described above. As shown in Tables 6.10 and 6.11, the results obtained were the correlation coefficients for each axis, the cumulative contribution rates, and the category weights. Figure 6.2 shows the category arrangement using these category weights. From this arrangement, we can classify the samples into four groups. The first group is those who have no knowledge of the product and rely on the prestige of department stores or brand names (*prestige group*). The second group is those who buy at supermarkets, regardless of the product (*indifferent group*). The third group is those who buy primarily brand-name products at speciality shops that have abundant knowledge about them (*high quality individual group*). The fourth group is those who rely on their own knowledge of products and will buy nonbrand-name products if they are to

Table 6.10. Correlation Coefficients and Cumulative Contribution Rates for the Survey on Young People's Purchasing Activities Survey

	Correlation Coefficient	Cumulative Contribution Rate
axis #1	6.13	50.4%
axis #2	4.98	83.7%

Table 6.11. Category Weights for the Young People's Purchasing Activities Survey

	Place of Purchase			Product Knowledge		Brand-Name Products	
	Supermarkets	Specialty Shops	Department Stores	None	Much	Doesn't Matter	Usually Buy
axis #1	0.337	−1.404	1.411	1.255	−0.616	0.311	0.560
axis #2	0.902	0.408	−0.253	0.503	−0.436	1.949	−1.286

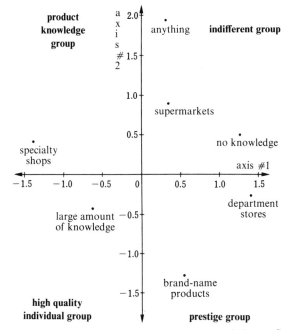

Fig. 6.2. Distribution of Categories in the Electric Razor Survey

their liking (*product knowledge group*). Samples 3, 4, and 6 can be classified as being in group 1 (*prestige group*), samples 1 and 7 in group 2 (*indifferent group*), sample 2 in group 3 (*high quality individual group*) and sample 5 in group 4 (*product knowledge group*).

6.5 FUZZY QUANTIFICATION THEORY IV

In Hayashi's Quantification Theory IV (*multi* dimensional scaling), we use numerical value $e_{ij}(i, j = 1, \ldots, n)$, which expresses the mutual intimacy among n individuals, as data. When numerical x_i is given for each individual i, we determine value x_i such that the relationship between the distance between samples i and j and the intimacy e_{ij} is monotonic. That is, we determine the one-dimensional arrangement of dividuals $i(i = 1, \ldots, n)$ that best realizes intimacy data e_{ij}. In Fuzzy Quantification Theory IV, the individual belongs to fuzzy group B, and the problem is to determine the above relationships.

Table 6.12. Data Handled by Fuzzy Quantification
Theory IV

			Individual		
			$1 \cdots i \cdots n$		
		Fuzzy Group B	$\mu_B(1) \cdots \mu_B(i) \cdots \mu_B(n)$		
Individual	1	$\mu_B(1)$	$e_{11} \cdots e_{1i} \cdots e_{1n}$		
	$\vdots$	$\vdots$	$\vdots \quad \vdots \quad \vdots$		
	i	$\mu_B(i)$	$e_{i1} \cdots e_{ii} \cdots e_{in}$		
	$\vdots$	$\vdots$	$\vdots \quad \vdots \quad \vdots$		
	n	$\mu_B(n)$	$e_{n1} \cdots e_{ni} \cdots e_{nn}$		

To put it another way, the effect of individuals with a small degree of belonging to fuzzy group B is not given much consideration. However, the effect of individuals with a large degree of attribution is given enough consideration, and numerical value x_i is determined. Table 6.12 shows the data handled in Fuzzy Quantification Theory IV.

Let us consider the unknown numerical value x_i given for the individual and amount Q, which is expressed using intimacy e_{ij}:

$$Q = -\sum_{i=1}^{n} \sum_{j=1}^{n} \mu_B(i)\mu_B(j)e_{ij}(x_i - x_j)^2.$$

Let us determine the one-dimensional arrangement $x_i(i = 1, \ldots, n)$. The origin and the scale of its dimension do not affect the relative relation among the individuals x_i, we will determine the arrangement x_i, which maximizes Q under conditions

$$\bar{x} = \sum_{i=1}^{n} \mu_B(i)x_i = 0$$

$$\sum_{i=1}^{n} x_i^2 = c$$

for convenience. Solving this using LaGrange's method of the undetermined multiplier, the arrangement x_i that maximizes Q can be written

$$\sum_{j=1}^{n} \mu_B(k)\mu_B(i)(e_{ki} + e_{ik})(x_i - x_k) = \lambda_{x_k}. \tag{6.20}$$

At this time, if we let

$$a_{ki} = (e_{ki} + e_{ik})\mu_B(k)\mu_B(i)$$

$$H = [h_{ij}]; \qquad h_{ij} = \begin{cases} a_{ij}; & i \neq j \\ -\sum\limits_{\substack{k=1 \\ k \neq i}} e_{ik}; & i = j \end{cases}$$

$$X = [x_1, \dots, x_n]',$$

Equation (6.20) gives the following eigenvalue problem:

$$HX = \lambda X$$

$$Q = X'HX = \lambda X'X = \lambda c.$$

Because of this, based on the condition

$$\sum_{i=1}^{n} x_i^2 = c$$

the x_i that maximizes Q is the eigenvector that corresponds to the maximum eigenvalue of H.

6.6 A NOTE ON APPLICATIONS

In this chapter we have explained fuzzy quantification theory using simple examples that are applications of fuzzy theory. We have described quantification methods, quantification of qualitative judgments that are expressed in terms of attribution, and, in particular, samples governed by the rules of fuzzy sets, following the concept of fuzzy events within the fuzzy set. In terms of characteristics, for example, when there is an order relation in a group that shows an external standard, quantification using Fuzzy Quantification Theory II must be carried out under order relation restrictions that preserve the order relation.[10] When the order relation can be expressed in terms of degree of attribution to the fuzzy set, this type of problem can be solved simply. For example, when we consider the case in which we have three groups, the "good group," the "bad group," and the "neither group," the problem can be solved by thinking of the samples in the "neither group" as being attributed 0.5 to the "good group" and 0.5 to the "bad group."

Fuzzy Quantification Theory II, which was developed early, is being used for the analysis of many actual problems: for example, in the primary factor analysis of Watada *et al.*,[11] and in the analysis of structural safety by Shiraishi *et al.*[12,13] and Watada[6]. We could not fit this discussion into our limited introduction in this chapter, but Watada *et al.*[14] have proposed a version of Fuzzy Quantification Theory I in which the values for object variablles and for categories are given by fuzzy numbers.

REFERENCES

(1) Hayashi, C., "On the Quantification of Qualitative Data from the Mathematico-Statistical Point of View," *Annals of the Institute of Stat. Math.*, **II**, 1 (1950). *See also* Hayashi, C., "On the Prediction of Phenomena from Qualitative Data and the Quantification of Qualitative Data from the Mathematico-Statistical Point of View," *Annals of the Institute of Stat. Math.*, 3, (1952).

(2) Hayashi, C., *Quantification Method*, Toyo Keizai Shinpo-sha (1972) (in Japanese). *See also* Hayashi, C., Higuchi, I., and Komazawa, T., *Statistics in Information Processing*, Sangyo Tosho (1970) (in Japanese).

(3) Dubois, D. and Prade, H., *Fuzzy Sets and Systems: Theory and Applications*, Academic Press (1980).

(4) Zadeh, L. A., "Fuzzy Sets," *Information and Control*, **8**, pp. 338–353 (1965).

(5) Zadeh, L. A., Probability Measures of Fuzzy Events, *Journal of Mathematical Analysis and Applications*, **23**, pp. 421–427 (1968).

(6) Watada, J., Fu, K. S., and Yao, J.T.P., "Damage Assessment Using Fuzzy Multi-variant Analysis," Technical Report No. CE-STR-84-4, School of Civil Engineering, Purdue University (1984).

(7) Watada, J., Tanaka, E., and Asai, K., "Fuzzy Quantification Type II," *The Japanese Journal of Behaviormetrics* (Behaviormetric Society of Japan), **9**, 2, pp. 24–32 (in Japanese).

(8) Tenenhaus, M., and Young, F. W., "Analysis and Synthesis of Multiple Correspondence Analysis, Optimal Scanning, Dual Scanning, Homogeneity Analysis and Other Methods for Quantifying Categorical Multivariate Data," *Psychometrica*, **50**, 1, pp. 91–119 (1985).

(9) Lebart, L., Morineau, A., and Warwick, K. M., *Multivariate, Descriptive Statistical Analysis*, John Wiley & Sons, New York (1984).

(10) Tanaka, Y., "An Application of Methods of Quantification to Analyze the Effects of Qualitative Factors," in Matsushita, K. ed., *Recent Developments in Statistical Inference and Data Analysis*, North-Holland (1980).

(11) Watada, J., Tanaka, E., and Asai, K., "Analysis of Purchasing Factors Using Fuzzy Quantification Theory Type II," *Journal of the Japan Industrial Management Association*, **32**, 5, pp. 51–65 (1981) (in Japanese).

(12) Shiraishi, N., Furuta, H., and Hashimoto, M., "Application of Fuzzy Quantification to Safety Evaluation of Structures," *Proceedings of 30th Structural Engineering Symposium*, pp. 227–284 (1984) (in Japanese).

(13) Shiraishi, N., Furuta, H., and Hashimoto, M., "Safety Evaluation of Structures Using Fuzzy Multicriterion Analysis," *Systems and Control*, **28**, 7, pp. 475–483 (1984) (in Japanese).

(14) Watada J., Tanaka, E., and Asai, K., "Fuzzy Quantification Theory Type I," *The Japanese Journal of Behaviormetrics* (Behaviormetric Society of Japan), **11**, 1, pp. 66–73 (1984).

Chapter 7

FUZZY
MATHEMATICAL
PROGRAMMING

In mathematical programming there are problems in which a real problem is described in terms of a mathematical model (model building) and problems in which an optimal solution is found from the model obtained (model solving). If a model approaches the actual problem without any limitations, it is said that the model becomes complicated and finding a solution is difficult; contradictory results are common. In addition, real problems include constraints and goals that are expressed in natural language. For example, the kind of ambiguity expressed in "we want to gain about A yen" or "we want to keep investments to about B yen or less" is included. Fuzzy mathematical programming, which expresses this kind of ambiguity in terms of fuzzy sets, comes close to our understanding. In this chapter we will discuss the formulation of fuzzy mathematical programming and its applications.

7.1   BASIC CONCEPT AND GENERAL FORMULATION

In conventional mathematical programming problems, we find a solution in which an objective function is maximized subject to constraints. In real problems we deal with things in which the constraints and objective functions are flexible. For example, in a corporate investment problem, there is ambiguity in the set comprising the amount of money available for investment in the planning stages. In addition, it is better for most corporations to follow standards for a degree of satisfaction, which means a certain amount of profit is acceptable, rather than to maximize corporate profits. In other words, an approach that handles objective sets following standards for a degree of satisfaction is closer to real problems than maximizing an objective function. Actually, it is said that Dantzig's early models did not have objective functions.[1] Objective functions are not essential; they are introduced simply to limit the solutions to one. The ambiguity of coefficients and its relationship to solutions for LP (Linear Programming) in real problems is discussed in reference (1).

From this point of view let us now talk about the fuzzy decisions that are obtained when fuzzy constraints and fuzzy objectives are given. Let the constraints and objectives be expressed by fuzzy sets C and G, and the membership functions for each be $\mu_C(x)$ and $\mu_G(x)$, respectively. In this case, fuzzy decision set D is defined as

$$D = C \cap G; \qquad \mu_D(x) = \mu_C(x) \wedge \mu_G(x), \tag{7.1}$$

meaning that it satisfies both the constraints and the objectives, where $\wedge$ means min. The membership function of fuzzy decision set D, $\mu_D(x)$, expresses the degree of belonging to decision set D. If $\mu_D(x) \leqq \mu_D(x')$, x' is better decision than x. Therefore, it can be considered appropriate to select x^* such that

$$\max_x \mu_D(x) = \max_x \mu_C(x) \wedge \mu_G(x) = \mu_C(x^*) \wedge \mu_G(x^*) \tag{7.2}$$

is the maximized solution.[2]

The maximized decision is defined by max and min operations, so let us consider the meaning. Assuming that there are two fuzzy sets $\mu_1(x)$ and $\mu_2(x)$, the functions that combine μ_1 and μ_2 with "and" and "or" are denoted as f, g: $[0, 1] \times [0, 1] \to [0, 1]$. More specifically,

$$\left. \begin{array}{l} (\mu_1 \text{ and } \mu_2)(x) = f(\mu_1(x), \mu_2(x)) \\[2mm] (\mu_1 \text{ or } \mu_2)(x) = g(\mu_1(x), \mu_2(x)) \end{array} \right\}. \tag{7.3}$$

For simplicity, let $\mu_1(x) = \alpha$ and $\mu_2(x) = \beta$; the functions are expressed $f(\alpha, \beta)$ and $g(\alpha, \beta)$. We assume the following axioms for f and g.

(1) f and g are continuous and nondecreasing.
(2) f and g are symmetrical for α and β, $f(\alpha, \beta) = f(\beta, \alpha)$ and $g(\alpha, \beta) = g(\beta, \alpha)$.

(3) $f(\alpha, \alpha)$ and $g(\beta, \beta)$ are strictly increasing functions.

(4) $f(\alpha, \beta) \leq \min[\alpha, \beta]$ and $g(\alpha, \beta) \geq \max[\alpha, \beta]$.

(5) $f(1, 1) = 1$ and $g(0, 0) = 0$.

(6) Logically equivalent fuzzy sets agree with each other. For example,

$$(\mu_1 \text{ and } (\mu_2 \text{ or } \mu_3))(x) = ((\mu_1 \text{ and } \mu_2) \text{ or } (\mu_1 \text{ and } \mu_3))(x).$$

Theorem 7.1. *Functions f and g that satisfy axioms 1–6 are limited to*

$$f(\mu_1, \mu_2) = \mu_1 \wedge \mu_2, \qquad g(\mu_1, \mu_2) = \mu_1 \vee \mu_2.$$

The proof of Theorem 7.1 is given in Reference (3). An "and" combination of fuzzy constraints and fuzzy objectives can be seen as the min, and if there is an "or" type combination within the fuzzy constraints, the combination is done with a max operation.

Let us now talk about the maximization problem in (7.2), which can be transformed into the following maximization problem using α-level sets:[4]

$$\sup_x \mu_D(x) = \sup_{\alpha \in [0,1]} [\alpha \wedge \sup_{x \in C_\alpha} \mu_G(x)], \qquad (7.4)$$

where $C_\alpha = \{x \mid \mu_C(x) \geq \alpha\}$.

If $\alpha_1 \leq \alpha_2$, we get

$$\sup_{x \in C_{\alpha 1}} \mu_G(x) \geq \sup_{x \in C_{\alpha 2}} \mu_G(x),$$

because $C_{\alpha 1} \supset C_{\alpha 2}$; therefore, the meaning of Equation (7.4) is easily understood. Setting

$$\sup_{x \in C_\alpha} \mu_G(x) = \phi(\alpha),$$

if the function $\phi(\alpha)$ is continuous for α, the optimal α^* is obtainable from $\alpha^* = \phi(\alpha^*)$. If $\phi(\alpha)$ is then continuous, we get the following optimization problem:

$$\sup_x \mu_G(x), \qquad \mu_C(x) \geq \mu_G(x) \qquad (7.5)$$

for (7.4), and we have standard mathematical programming that does not include the logical operation $\wedge$. If fuzzy set C is a strongly convex fuzzy set, $\phi(\alpha)$ is continuous. Fuzzy set C is strongly convex if and only if the following holds:

$$\mu_C(\lambda x + (1 - \lambda)y) > \mu_C(x) \wedge \mu_C(y); \qquad {}^\forall \lambda \in [0, 1]. \qquad (7.6)$$

MILLS COLLEGE
LIBRARY

If function $\phi(\alpha)$ is continuous, the fuzzy mathematical programming problem can be solved by finding α^* such that

$$\alpha = \sup_{x \in C_\alpha} \mu_G(x).$$

Therefore, the solution can be obtained using the following algorithm.

Algorithm

(1) $\forall \partial_k$ is established for $k = 1$.
(2) Calculate

$$\mu_G^k(x^*) = \max_{x \in C_{\alpha_k}} \mu_G(x).$$

(3) If $|\varepsilon_k| = |\alpha_k - \mu_G^k(x^*)| > \varepsilon$, go to (4), and if $|\varepsilon_k| \leqq \varepsilon$, go to (5).
(4) Let $\alpha_{k+1} = a_k - r_k \varepsilon_k$ and $k = k + 1$ and go to (2), where r_k such that $0 \leqq \alpha_{k+1} \leqq 1$ is chosen.
(5) Letting $\alpha^* = \alpha_k$, find x^* such that

$$\max_{x \in C_{\alpha_k}} \mu_G(x) = \mu_G(x^*),$$

and end this process.

In the algorithm above, step (2) is finding the solution by means of standard mathematical programming.

7.2   FUZZY LINEAR PROGRAMMING

When the system and fuzzy coefficients are linear, it is called a fuzzy LP problem. Fuzzy linear programming problems have been formulated from various points of view, and here we will describe three different formulations.

7.2.1   Fuzzy LP Problems Using Fuzzy Inequalities

We will describe the fuzzy LP formulation of H. J. Zimmermann.[5] The objectives and constraints are given by the following fuzzy inequalities:

$$c^t x \lesssim z, \qquad Ax \lesssim b, \qquad x \geq 0, \tag{7.7}$$

where $\lesssim$ is a fuzzy inequality, and the meaning of the equations in (7.7) is that we have a problem in which constraint Ax is about b or less and outlay

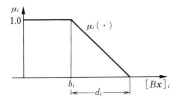

Fig. 7.1. Example of Fuzzy Set "About b_i or less"

$c^t x$ is about z or less. An example of the membership function for the fuzzy inequality about b_i or less is shown in Fig. 7.1.

Since the objectives and constraints are expressed by inequalities, we can consolidate and write $Bx \lesssim b$. The ith inequality about b_i or less is defined by the following membership functions:

$$\left.\begin{array}{ll} \mu_i([Bx]_i) = 1; & [Bx]_i \le b_i \\ 0 \le \mu_i([Bx]_i) \le 1; & b_i \le [Bx]_i \le b_i + d_i \\ \mu_i([Bx]_i) = 0; & [Bx]_i \ge b_i + d_i \end{array}\right\}, \qquad (7.8)$$

where $[Bx]_i$ is the ith element of the vector, μ_i the membership function of the ith inequality, and d_i the maximum possible value for the right-hand side of the inequality.

The problem of the maximized decision for (7.2) is finding x such that

$$\max_{x \ge 0} \min_i \{\mu_i([Bx]_i)\}. \qquad (7.9)$$

Let

$$\mu_i([Bx]_i) = \begin{cases} 1; & [Bx]_i \le b_i \\ 1 - \dfrac{[Bx]_i - b_i}{d_i} & b_i \le [Bx]_i \le b_i + d_i \\ 0; & [Bx]_i \ge b_i + d_i \end{cases} \qquad (7.10)$$

be the linear constraints, as in Fig. 7.1. We do the normalization $b_i' = b_i/d_i$, $[B']_{ij} = [B]_{ij}/d_i$, and considering that the constraints are linear, (7.9) comes out as

$$\max_{x \ge 0} \min \{b_i' - [B'x]_i\} \qquad (7.11)$$

(7.11) turns out to be the following standard LP problem:

$$\left.\begin{array}{l} \max \lambda \\ \lambda \le b_i' - [B'x]_i; \qquad i = 1,\dots,m \end{array}\right\}. \qquad (7.12)$$

In this formulation, the solutions to fuzzy LP problems can be obtained using standard LP.

Example 7.1 (H.-J. Zimmermann[5]).   We will use a real transport problem involving four types of trucks, x_1–x_4 to explain fuzzy LP problems. First, consider the following standard LP problem in comparison with the fuzzy LP problem:

$$\min_{x} z = 41\,400x_1 + 44\,300x_2 + 48\,100x_3 + 49\,100x_4 \left.\begin{array}{l} \\ 0.84x_1 + \quad 1.44x_2 + \quad 2.16x_3 + \quad 2.40x_4 \geq 170 \\ 16x_1 + \quad\; 16x_2 + \quad\;\; 16x_3 + \quad\;\; 16x_4 \geq 1300 \\ x_1 \qquad\qquad\qquad\qquad\qquad\qquad\quad\; \geq 6 \end{array}\right\}. \quad 7.13)$$

(7.13) is given in clear numerical values. The solution to this LP problem is

$$x_1^* = 6, \quad x_2^* = 17.85, \quad x_3^* = 0, \quad x_4^* = 58.64, \quad z^* = 3918850. \quad (7.14)$$

Now let the right side of the constraints be ambiguous and the data in Table 7.1 be given. The membership functions of the fuzzy constraints in Table 7.1 are given by a straight line from $\mu = 0$ to $\mu = 1$. When we put this LP problem into the form of (7.12), we get

$$\max \lambda$$

$$\left.\begin{array}{l} \lambda \leq \quad 8.4 - 0.083x_1 - 0.089x_2 - 0.096x_3 - 0.098x_4 \\ \lambda \leq -17 + 0.084x_1 + 0.144x_2 + 0.216x_3 + 0.24\; x_4 \\ \lambda \leq -14 + 0.16\; x_1 + 0.16\; x_2 + 0.16\; x_3 + 0.16\; x_4 \\ \lambda \leq \;-1 + 0.167x_1, \qquad x \geq 0 \end{array}\right\}, \quad (7.15)$$

and the solution to the problem is shown in Table 7.2.

Table 7.1.   Non-Fuzzy Constraints and Fuzzy Constraints

	Non-Fuzzy Constraints	Fuzzy Constraints	
		$\mu = 0$	$\mu = 1$
objective function		4 200 000	3 700 000
1st constraint	170	170	180
2nd constraint	1 300	1 300	1 400
3rd constraint	6	6	12

Table 7.2. Solutions for Non-Fuzzy
LP Problems and Fuzzy LP

Standard Solution	Fuzzy Solution
$x_1 = 6$	$x_1 = 17.41$
$x_2 = 17.85$	$x_4 = 66.54$
$x_4 = 58.65$	
$z = 3\,918\,850$	$z = 3\,988\,257$
value for constraint	value for constraint
1. 171.5	1. 174.2
2. 1 300	2. 1 342.4
3. 6	3. 17.4

In a fuzzy LP problem, the conditions are considered strictly as in letting the first constraint be "about 170 or more," $\mu_1(170) = 0$. In addition, instead of an objective function, we select a permissible range such as "from about 3,700K to about 4,200K." Since a fuzzy LP problem can be constructed from fuzzy information in this way, it is not necessary to stipulate the LP problem clearly in advance at a great cost. However, the costs for the fuzzy LP problem are 1.7% higher than those for the standard LP problem.

7.2.2 Fuzzy LP Problems Using the Linear Interval Method

Here we will describe Negoita's formulation of fuzzy LP problems.[6]

The features of this type of problem are that the ambiguity of the coefficients of the linear constraints is expressed as a fuzzy set, and it uses the concepts of interval programming.

Let us consider the following LP problem:

$$\left. \begin{array}{l} \max c^t x \\[2mm] K_1 x_1 + \cdots + K_n x_n \subset K, \qquad x \geq 0 \end{array} \right\}, \tag{7.16}$$

where $K, K_1, \ldots, K_n$ are convex fuzzy sets. Convex fuzzy sets are defined by (7.6) using the equality. If we use the concept of level sets, this problem can be transformed into a linear interval programming problem. If we now let fuzzy set K be made up of r number of level-sets, K is expressed by

$$K = \alpha_1 \cdot [K]_{\alpha 1} + \cdots + \alpha_r \cdot [K]_{\alpha r}, \tag{7.17}$$

where the $+$ sign means $\cup$. If we introduce the level-sets, (7.16) gives the following linear interval programming problem:

$$\left.\begin{array}{l} \max c^t x \\ x_1 \cdot [K_1]_{\alpha i} + \cdots + x_n \cdot [K_n]_{\alpha i} \subset [K]_{\alpha i}; \qquad i = 1, \ldots, r, \qquad x \geq 0 \end{array}\right\}. \quad (7.18)$$

The following is a numerical example for the α_i level-set in (7.18).

Example 7.2 (Negoita, et al.[6]). Let us consider the following LP problem in which the coefficients are expressed by intervals:

$$\left.\begin{array}{l} \min 3x_1 + 2x_2 \\ [0.02, 0.03]x_1 + [0.05, 0.06]x_2 \subseteq [10.5, 15.6] \\ [0.1, 0.5]x_1 - [0.01, 0.02]x_2 \subseteq [0.35, 0.53]; \qquad x \geq 0 \end{array}\right\}. \quad (7.19)$$

From the point of view of guaranteeing that, even in the worst case values for coefficients, solution x is within the interval of the constraints, we can exchange (7.19) for the following LP problem:

$$\left.\begin{array}{l} \min 3x_1 + 2x_2; \qquad x \geq 0 \\ 0.03x_1 + 0.06x_2 \leq 15.6, 0.5x_1 - 0.01x_2 \leq 0.53 \\ 0.02x_1 + 0.05x_2 \geq 10.5, 0.1x_1 - 0.02x_2 \geq 0.35 \end{array}\right\}. \quad (7.20)$$

In the fuzzy LP of (7.18), the fuzzy constraints are exchanged for $2r$ number of standard constraints. The number of constraint equations gets quite large, but it is impossible to avoid this kind of complication when dealing with flexible constraints.

7.2.3 Possibilistic Linear Programming Problems [7]

Here we will deal with the case in which we know only that the coefficients for the LP problem are ambiguous, and we will look at this ambiguity in terms of fuzzy numbers. The fuzzy numbers are given as information from experts, and we will consider their possibility distributions. We will consider fuzzy inequalities with fuzzy coefficients.

$$\left.\begin{array}{ll} A_{11}x_1 + \cdots + A_{1n}x_n \gtrsim B_1 & \text{fuzzy goal} \\ A_{21}x_1 + \cdots + A_{2n}x_n \lesssim B_2 & \text{fuzzy constraint} \\ \vdots \qquad \vdots \quad \vdots \qquad \vdots \\ A_{m1}x_1 + \cdots + A_{mn}x_n \lesssim B_m & \text{fuzzy constraint} \end{array}\right\}. \quad (7.21)$$

If we do not separate the objectives and constraints, we can write (7.21) as

$$Y_i = A_{i0}x_0 + A_{i1}x_1 + \cdots + A_{in}x_n \gtrsim 0; \qquad i = 1,\ldots,m, \qquad (7.22)$$

where $A_{i0} = B_i$ and $x_0 = 1$, and A_{ij} includes the sign of the fuzzy set.

As was mentioned in Chapter 4, (7.22) can be considered to be a possibilistic linear system, and if we let the coefficients be L fuzzy numbers, the membership function of Y_i can be obtained from Theorem 4.1 as

$$\mu_Y(y) = L((y - x^t\boldsymbol{\alpha})/c^t|x|), \qquad (7.23)$$

where the fuzzy coefficient is $A_i = (\alpha_i, c_i)$.

Since we can easily calculate the fuzzy set for Y_i, we must define $Y_i \gtrsim 0$. According to Dubois and Prade,[8] the definitions of inequalities for fuzzy numbers are divided into 4 types, but for simplicity we make the following definition here:

$$Y_h \gtrsim 0 \leftrightarrow \mu_{Y_i}(0) \leq 1 - h, \qquad \alpha^t x \geq 0, \qquad (7.24)$$

where h shows the degree to which Y_i is $\gtrsim 0$, and the larger the h, the stronger the meaning of "about positive" (see Fig. 7.2).

If, for simplicity, we let the fuzzy coefficients be triangular, we get

$$\mu_{Y_i}(y) = 1 - \frac{|y - x^t\alpha|}{c^t|x|} \qquad (7.25)$$

for (7.23) (see Chapter 4.) Given uniformly in vector form, (7.22) is

$$Y = A_i x \gtrsim 0, \qquad (7.26)$$

where we let $A_i = (\alpha_i, c_i)$. (7.24) turns out to be

$$\mu_{Y_i}(0) = 1 - \frac{\alpha_i^t x}{c_i^t|x|} \leq 1 - h, \qquad \alpha_i^t x \geq 0, \qquad (7.27)$$

and since $x \geqq 0$ is fixed, (7.27) comes out simply as follows:

$$(\alpha_i - hc_i)^t x \geq 0; \qquad i = 1,\ldots,m. \qquad (7.28)$$

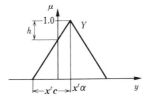

Fig. 7.2. Fuzzy Inequality $Y \gtrsim 0$

As was mentioned in 7.1, finding x such that the maximum value for degree h satisfies m number of inequalities is the fuzzy LP problem we are dealing with here. In other words, we need to find h^* and x^* such that

$$\max_{h,x} h = h^*, \tag{7.29}$$

subject to the constraints in (7.28). This is a nonlinear programming problem, but in this formulation the solution can be obtained by considering the ambiguity of all of the coefficients simultaneously.

From a different point of view, the above problem can be transformed into a programming problem for finding the possibility of a solution. In (7.21), let us say that the term B_i is the only ambiguous one, and let us consider the problem of finding the possibility of a solution in terms of fuzzy number X_i. In other words,

$$Y_i = B_i + a_{i1}X_1 + \cdots + a_{in}X_n \gtrsim 0; \qquad i = 1,\ldots,m, \tag{7.30}$$

where X_i is the fuzzy solution and is expressed by $X_i = (x_i, c_i)$. Let $B_i = (b_i, d_i)$, and let fuzzy numbers X_i and B_i be triangular. Given h, the fuzzy inequality can be reduced to

$$\sum_j (a_{ij}x_j - h_i|a_{ij}|c_j) + b_i - h_i d_i \geq 0. \tag{7.31}$$

The problem here is finding out the possibility of X_i, which reflects the possibility of B_i. Therefore, by means of the width c_i of solution X_i, we introduce the performance index

$$J = k_1 c_1 + \cdots + k_n c_n \tag{7.32}$$

as a measure of the possibility of solution, where $k_i \geq 0$ is the weight coefficient that tells which solution X_i we want to make larger.

This problem means finding the fuzzy solution $X = (x, c)$ that maximizes J in (7.32), subject to the constraints in (7.31). In order to explain the meaning of the problem, let us consider the following example.

Example 7.3 (Tanaka, et al.[7]).   Fuzzy constraints are given as follows:

$$\left.\begin{array}{ccc} X_1 + X_2 \gtrsim \underline{6}, & 2X_1 + X_2 \lesssim 12, & X_2 \lesssim \underline{5} \\ 3X_1 + 4X_2 \lesssim 28, & X_1 \lesssim \underline{6}' \end{array}\right\}, \tag{7.33}$$

where $\underline{a}$ is a fuzzy number and $\underline{6} = (6, 0.6)$, $12 = (12, 1.8)$, $\underline{5} = (5, 1.0)$, $\underline{28} = (28, 2.8)$, and $\underline{6}' = (6, 0.9)$. Here we let the idea be that the decision maker wants to make the possibility of x_1 large. In other words, since the circumstances are ambiguous, he or she wants to know the maximum possibility for X_1. Therefore, we let the objective function be

$$J = c_1. \tag{7.34}$$

The fuzzy inequalities in (7.33) come out as

$$\left. \begin{array}{l} x_1 \;\; + x_2 - h(c_1 + c_2) \;\;\;\;\; + 6 - 0.6h \geq 0 \\ -2x_1 \;\; - x_2 - h(2c_1 + c_2) \;\; + 12 - 1.8h \geq 0 \\ \;\;\;\;\;\;\;\;\;\; -x_2 - hc_2 \;\;\;\;\;\;\;\;\;\;\;\;\;\; + 5 - \;\;\;\; h \geq 0 \\ -3x_1 - 4x_2 - h(3c_1 + 4c_2) + 28 - 2.8h \geq 0 \\ -x_1 \;\;\;\;\;\;\;\;\;\; - hc_1 \;\;\;\;\;\;\;\;\;\;\; + 6 - 0.9h \geq 0 \end{array} \right\}, \tag{7.35}$$

and the solution X^* that maximizes J for $h = 0.5$ is

$$X_1^* = (2.94, 1.24), \qquad X_2^* = (3.98, 0). \tag{7.36}$$

In solution X_2^*, there is no ambiguity because $c_2 = 0$, and all of the ambiguity in B_i is brought together in X_1^*. If we wish to consider solutions when degree $h = 0.5$ or more, the possibility interval for x_1 is $[2.32, 3.56]$. With this interval, the decision maker can consider conditions that are not incorporated into the mathematical model and make his decision.

7.3 SUPPLEMENTARY NOTE

Since, in the fuzzy mathematical programming described in this chapter, constraints and objectives are not divided up and are handled using identical concepts, multiple-objective programming problems can be considered. There are various studies concerning the unification of multiple objectives, and unification by sum $\Sigma\mu_i(x)$ and product $\Pi\mu_i(x)$, as well as unification by the more general fuzzy integration,[9] can be considered. In addition, ideas such as fuzzy multiple-objective programming problems using dialogue methods[10] and fuzzy multiple-objective programming problems in which fuzzy superiority structures are introduced[11] are being dealt with. Since decision making problems are problems that people participate in, various fuzzy multiple-objective programming problems can be formulated by introducing fuzzy sets. Furthermore, D. Dubois has uniformly formulated various types of fuzzy LP problems and has uniformly described fuzzy LP problems from measures of possibility and necessity.[12] W. J. M. Kickert has a book dedicated to fuzzy mathematical programming,[13] and Chapter 11 of Sakawa's book explains fuzzy linear programming.[14]

REFERENCES

(1) Moriguchi, S., "Discussions with Dr. Dantzig," *Journal of Keiei Kagaku*, **3**, 3, pp. 166–168 (1960) (in Japanese).

(2) Bellman, R. E., and Zadeh, L. A., "Decision Making in a Fuzzy Environment," *Management Science*, **17**, pp. 8141–8164 (1970).

(3) Bellman, R. E. and Giertz, M., "On the Analytic Formalism of the Theory of Fuzzy Sets," *Information Sciences*, **5**, pp. 149–156 (1973).

(4) Tanaka, H., Okuda, T., and Asai, K., "On Fuzzy Mathematical Programming," *Journal of Cybernetics*, **3**, pp. 37–46 (1974).

(5) Zimmermann, H. J., "Description and Optimization of Fuzzy Systems," *International Journal of General Systems*, **2**, pp. 209–215 (1976).

(6) Negoita, C. V., and Sularia, M., "On Fuzzy Mathematical Programming and Tolerances in Planning," *ECECSR Journal*, **1**, pp. 3–14 (1976).

(7) Tanaka, H., and Asai, K., Fuzzy Solutions in Fuzzy Linear Programming Problems," *IEEE Trans. on SMC*, **14**, pp. 325–328 (1984).

(8) Dubois, D., and Prade, H., "Ranking Fuzzy Numbers in the setting of Possibility Theory, "*Information Sciences*, **30**, pp. 183–224.

(9) Ichihashi, H., Tanaka, H., and Asai, K., "Fuzzy Integrals and Their Applications to Multiple Attribute Decision Problems," *Keisoku Jidou Seigyo Gakkai*, **22**, 5, pp. 557–562 (1986) (in Japanese).

(10) Sakawa, M., Yano, H., and Yumine, T., "An Interactive Fuzzy Satisfying Method for Multiobjective Linear-Programming Problems and Its Application," *IEEE Trans. on Systems, Man, and Cybernetics*, SMC-17, pp. 654–661.

(11) Takeda, E., and Nishida, T., "Multiple Objective Linear Program Problems with Fuzzy Domination Structures," *Fuzzy Sets and Systems*, **3**, pp. 123–136 (1980).

(12) Dubois, D., Linear Programming with Fuzzy Data, in Bezdek, J. C., ed., *The Analysis of Fuzzy Information*, **3**, CRC Press, Florida (1987).

(13) Kickert, W. J. M., *Fuzzy Theories on Decision Making*, Martinus Nijjhoff Social Sciences Division (1978) Leiden.

(14) Sakawa, M., *Optimization of Linear Systems*, Morikita Shuppan (1984) Tokyo.

Chapter 8

EVALUATION

Problems involving evaluation of objects with several attributes are an important field of application for fuzzy theory. Evaluation problems have an especially close relationship with decision making problems, and one could say that "decision making" is another way of saying "evaluation." However, we will only deal with evaluation that does not involve conscious decision making.

The idea of using the concept of fuzziness in evaluation problems comes from ambiguity viewed from the following two points of view. First is the ambiguity in the characteristics of the object of evaluation. For example, if we think about evaluating the quality of houses, we understand that such characteristics of a house as the quality of the environment or what features are built into the house are ambiguous, even if quantified. Second is the ambiguity of the evaluation measure for the subject being evaluated. Even if we temporarily say that the object has no ambiguity and its attributes are strictly quantified, the method of obtaining those values cannot be anything but subjective. In this way two types of ambiguity, that of the object of the evaluation and that in the subject evaluating, exist in evaluation problems. In this chapter we will deal with ambiguity in the measure of the subject doing the evaluation. We will make a model of the evaluation process using this measure and consider the identification of the evaluation structure.

We will use fuzzy measures and fuzzy integrals, which are necessary tools in fuzzy theory. *Fuzzy measures* are criteria for measuring attributes of objects such as length and area, and they are a loosening of the conditions imposed by standard criteria.

8.1   FUZZY MEASURE

When we consider a certain set X, the function g that makes subsets E and F correspond to the values in the interval $[0, 1]$ are called fuzzy measures if they have the following properties:

$$(1)\ \ g(\phi) = 0, \qquad g(X) = 1 \tag{8.1}$$

$$(2)\ \ \text{If } E \subset F, \qquad g(E) \leqq g(F) \tag{8.2}$$

$$(3)\ \ \text{If } E_1 \subset E_2 \subset \cdots \text{ or } E_1 \supset E_2 \supset \cdots \lim_{n \to \infty} g(E_n) = g\left(\lim_{n \to \infty} E_n\right) \tag{8.3}$$

Property (1) expresses boundedness, property (2) monotony, and property (3) continuousness. In the following, for simplicity, let X be a finite set. In this case, the third condition is unnecessary. As with fuzzy measures, the functions that make sets correspond to numerical values are called set functions. Fuzzy measures are monotonic set functions. Condition (2) is the same as the following condition.

$$(2)'\ \ g(G \cup H) \geq g(G) \vee g(H). \tag{8.4}$$

If the equal sign arises in $(2)'$, g is referred to as a possibility measure.

Now let us think about standard measures. If we express a half-open interval in real number line R as $(a, b]$, the length of the interval can be expressed by function μ:

$$\mu((a, b]) = b - a.$$

Since the interval $(a, b] = \{x \mid a < x \leqq b\}$ can be seen as a subset of R, it is appropriate to think of μ as a set function. If we let $a < b < c$, this clearly gives rise to

$$\mu((a, c]) = \mu((a, b]) + \mu((b, c]).$$

Here, $(a, b]$ and $(b, c]$ have no point in common, and since interval $(a, c]$ can be expressed as the union of $(a, b]$ and $(b, c]$, as in

$$(a, c] = (a, b] \cup (b, c],$$

μ has properties such that

$$\mu((a, b] \cup (b, c]) = \mu((a, b]) + \mu((b, c]).$$

Expressing this in general terms, when $E \cap F = \phi$, we get

$$\mu(E \cup F) = \mu(E) + \mu(F), \tag{8.5}$$

and this is called *additivity*, that is, it shows the properties of the measure. All standard measures have additivity. As an example other than length, we can consider probability. Let X be the set of numbers on dice. If we let the event in which a 1 or a 2 appears when a die is rolled be expressed by the set $\{1, 2\}$, the probability can be expressed using set function P such that $P(\{1, 2\}) = 1/3$. P can be called a probability measure. P is clearly additive. For example, we get

$$P(\{1, 2\} \cup \{3, 4\}) = P(\{1, 2\}) + P(\{3, 4\}).$$

Since additive measures are monotonic, they are one type of fuzzy measure. Conversely, we can say that fuzzy measures are generalizations of additive measures. If, when $E \cap F = \phi$,

$$g(E \cup F) \geq g(E) + g(F), \tag{8.6}$$

g is said to show *superior additivity*, and if

$$g(E \cup F) \leq g(E) + g(F), \tag{8.7}$$

g is said to show *sub additivity*. Fuzzy measures can be interpreted as subjectivity measures of the people who measures objects, but what are called measures of human subjectivity are not additive in most cases. Additivity is a strict condition. For example, take the concept of subjective probability; given the condition that this shows additivity, we have a measure for subjective probability evaluation. As an example of a nonadditive measure we can consider a standard measure that includes utility. Let us look at length. Say that we have poles of lengths 1, 2, and 3, and let the measure that includes their utility be g. We can easily see that, depending on the person, we might get

$$g(\text{length} = 3) \geqq g(\text{length} = 1) + g(\text{length} = 2)$$

or

$$g(\text{length} = 3) < g(\text{length} = 1) + g(\text{length} = 2).$$

If we consider the measures for attributes of objects of evaluation and the quantification of the attributes of those objects, we can evaluate complete objects. The evaluation models that we will deal with here use fuzzy measures for evaluation measures, and express the overall evaluation values for objects as fuzzy integrals of quantified attributes.

8.2   FUZZY INTEGRALS

The fuzzy integral of function $h: X \to [0, 1]$ on $E (\subset X)$ by fuzzy measure g is defined as follows:

$$f_E h(x) \circ g = \sup_{\alpha \in [0,1]} [\alpha \wedge g(E \cap H_\alpha)];$$

$$H_\alpha = \{x \mid h(x) \geqq \alpha\}. \tag{8.8}$$

E is what is called the *domain of integration*, and when $E = X$, it is written in abbreviated form (in this case, $X \cap H_\alpha = H_\alpha$). H_α gets smaller as α gets larger, and since g is monotonic the value for $g(E \cap H_\alpha)$ decreases as α gets larger. Fuzzy integrals have the following properties:

(1) $fa \circ g = a, a \in [0, 1].$ $\tag{8.9}$

(2) If $h_1 \leqq h_2$, $f h_1(x) \circ g \leqq f h_2(x) \circ g.$ $\tag{8.10}$

(3) If $E \subset F$, $f_E h(x) \circ g \leqq f_F h(x) \circ g.$ $\tag{8.11}$

The essential property of fuzzy integrals is its monotonic nature, which reflects the properties of fuzzy measures. When X is a finite set such that

$$X = \{x_1, x_2, \ldots, x_n\},$$

we let function h be

$$h(x_1) \geqq h(x_2) \geqq \cdots \geqq h(x_n).$$

When X is a finite set, we can think this way if we order any function according to size and renumber the elements of X accordingly. If we do so, the fuzzy integral is expressed by

$$f h(x) \circ g = \bigvee_{i=1}^{n} [h(x_i) \wedge g(H_i)]; \tag{8.12}$$

$$H_i = \{x_1, x_2, \ldots, x_i\}.$$

The results of the integration are as shown in Fig. 8.1. For calculating the fuzzy

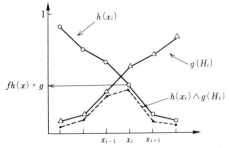

Fig. 8.1. Calculation of the Fuzzy Integration

integration here, knowing the value of g for n number of subsets $H_1, H_2, \ldots, H_n$ of X formed in correspondence with the form of function h is enough. Actually, it is convenient if we know the equation that gives the value of g for any given subset of X. For example, in the case of probability, if we let p^i be the probability density that corresponds to x_i, as in

$$P(H_i) = \sum_{k=1}^{i} p^k,$$

the probability of any set (event) is expressed using the probability of that element. Since this is derived from the condition of additivity, it is not always possible to do this for any fuzzy measure, but the following special fuzzy measure has been considered.

When $E \cap F = \phi$,

$$g_\lambda(E \cup F) = g_\lambda(E) + g_\lambda(F) + \lambda g_\lambda(E)g_\lambda(F); \qquad -1 < \lambda < \infty \quad (8.13)$$

λ is a parameter, and if $\lambda > 0$, g_λ shows superior additivity; if $\lambda = 0$, it shows additivity; and if $\lambda < 0$, it shows subadditivity.

If we let

$$g^i = g_\lambda(\{x_i\}), \tag{8.14}$$

we get

$$g_\lambda(H_i) = g^i + g_\lambda(H_{i-1}) + \lambda g^i g_\lambda(H_{i-1})$$
$$= \frac{1}{\lambda}\left[\prod_{k=1}^{i}(1 + \lambda g^k) - 1\right], \tag{8.15}$$

and we obtain a formula that is the same as that for probability. g^i is called the *density* of fuzzy measure. Fuzzy measure g_λ is extremely general, and models with this subjective measure are frequently used.

Now let us consider an evaluation model for a concrete example. The objects of the evaluation are houses, and we will construct a model for evaluating the desirability of houses. The attributes of the houses that we will deal with are x_1 = price, x_2 = size, x_3 = facilities, x_4 = place, and x_5 = living environment. Thus,

$$X = \{x_1, x_2, x_3, x_4, x_5\}$$

expresses the set of attributes of the houses. Let there be m number of houses, and let

$$h_j: X \to [0, 1]$$

be the function that expresses the evaluation value of the attributes of the jth house. For example, $h_j(x_1)$ is a value standardized within the interval $[0, 1]$, and $h_j(x_5)$ is the living environment quantified by some method and expressed in a numerical value between 0 and 1. We then let the fuzzy measure be the degree of consideration of an attribute during the desirability evaluation. For example, $g\{x_1\})$ is the degree to which we consider price, and $g(\{x_3, x_4\})$ is the degree to which we consider facilities and place, when we consider them simultaneously. Since we can see that the degree of consideration gets larger when the number of attributes increases, g can be interpreted as a fuzzy measure.

Under the above assumptions, the desirability of the jth house e_j is given by the fuzzy integral:

$$e_j = \int h_j \circ g. \tag{8.16}$$

This is a fuzzy integral model for evaluation. The problem here is identification of the model, in other words, identification of the evaluation structure and fuzzy measure. In general, when we can think of there being a large number of attributes, the evaluation structure involves finding out which of these attributes are necessary for evaluation. First, let us say that the structure is known and think about the identification of the fuzzy measure.

We consider the desirability of each of m number of houses for a single person and let it be d_j expressed as a numerical value from 0 to 1. For $1 \leq j \leq m$, we find a fuzzy measure that minimizes the difference between d_j and e_j. In this case all we have to do is standardize e_j so that the maximum and minimum values of d_j and e_j are equal to each other and find the difference. If the fuzzy measure is constructed from a density such as g_λ, the identification comes down to the problem of determining density $g_i(1 \leq i \leq 5)$ as a parameter; but in general the number of values for g that must be determined is that of the number of subsets of X. When we are dealing with this type of evaluation problem, what we must be most careful about is the independence of attributes and evaluation measures. Up to now, linear evaluation models have been the sum of ω_i weighted attribute evaluation values, as in

$$e_j = \sum_{i=1}^{5} \omega_i h_j(x_i), \qquad (8.17)$$

and it was allowed that this model was efficient when based on the assumption that the attributes and measures were independent. The meaning of independence of attributes is that no correlation between two attributes can be found. For example, it is difficult to say that price and size are independent. If the correlation is large, one of them can be thrown away, but there is the problem of what to do if the correlation is intermediate in nature. Even if the attributes are objectively independent, there are cases when the subjective measures are not necessarily so. Here the independence of measures means that they show additivity. For example, even if $x_3 =$ facilities and $x_5 =$ living environment are independent, the degree of consideration might be

$$g(\{x_3, x_5\}) \neq g(\{x_3\}) + g(\{x_5\}).$$

It is not necessary to assume that attributes and measures are independent in fuzzy integral models, and considering this point, we find they are more general than linear models. In addition, by inference from the point of view of the independence of fuzzy measures identified, we can solve evaluation structure identification problems. The identification of structure involves choosing the necessary attributes only, but using fuzzy measures, we can consider two clues. The first is the size of the density of the fuzzy measure $g^i = (\{x_i\})$. Since this is the degree of consideration of attribute x_i, an attribute with a degree of consideration that is much smaller than the others is unnecessary. The second is the distance from the additivity of the fuzzy measure. Let us temporarily assume that x_1 and x_2 represent exactly the same attribute. If so we might get

$$g(\{x_1, x_2\}) = g(\{x_1\}) = g(\{x_2\}).$$

Actually the part of the fuzzy integration related to x_1 and x_2 is

$$(h(x_1) \wedge g(\{x_1\})) \vee (h(x_2) \wedge g(\{x_1, x_2\})),$$

$h(x_1)$ and $h(x_2)$ are equal, and $g(\{x_1, x_2\})$ is larger than $g(\{x_1\})$; therefore, this term turns out to be

$$h(x_1) \wedge g(\{x_1, x_2\}).$$

Therefore, the size of $g(\{x_1\})$ can be anything below $g(\{x_1, x_2\})$. From this, when we are doing the identification calculations for g, we get the above results if we let $g\{\{x_1\})$ have its maximum value $g(\{x_1, x_2\})$. This is the same for $g(\{x_2\})$.

Rewriting the results with $g^1 = g^2$, we get

$$g(\{x_1, x_2\}) = g^1 \vee g^2.$$

Next, we let the attributes differ, but let their evaluation values have complete correspondence with each other. From the properties of fuzzy integration, we can see that we get

$$g(\{x_1, x_2\}) = g^1 \vee g^2.$$

In other words, when we have two attributes with a large degree of sub-additivity, one of them (the smaller of g^1 and g^2) is unnecesary. In addition, an attribute that does not increase the degree of consideration even if it has no large correspondence with another attribute can be thought of as unnecessary. In this way we obtain a clue to the fact that we have mistakenly included an unnecessary attribute and should remove it.

At this point, once again, we define the following amounts for

$$X = \{x_1, x_2, \ldots, x_n\}:$$

$$\mu_{ij} = \begin{cases} \dfrac{g(\{x_i, x_j\}) - (g^i + g^j)}{g^i \wedge g^j}; & i \neq j \\ 0; & i = j \end{cases} \qquad (8.18)$$

$$m_{ij} = \begin{cases} \mu_{ij}; & \mu_{ij} \leq 0 \\ \mu_{ij}/(\mu_{ij} + 1); & \mu_{ij} > 0 \end{cases} \qquad (8.19)$$

$$\eta_j = \sum_{i=1}^{n} m_{ij}^3/(n - 1). \qquad (8.20)$$

Here, μ_{ij} is a measure of the gap from additivity of g for x_1 and x_2. If we have superior additivity, $\mu_{ij} > 0$; if we have additivity, $\mu_{ij} = 0$. If

$$g(\{x_1, x_2\}) = g^1 \vee g^2,$$

we get μ_{ij}. m_{ij} is a standardization of μ_{ij} in values from -1 to 1, and η_j is a measure of the gap between the degree of additivity of attribute x_j and the average of that of the other attributes. When attribute x_j has even more inferior additivity than the other attributes, we can think of it as meaning that x_j is an overlapping attribute. We can call η_j the degree of overlap. A degree of overlap near -1 is a high degree of overlap. Now let us consider the size of both consideration degree g^j and degree of overlap η_j and make the following definition:

$$\xi_j = 1 + \eta_j(1 - g^j). \tag{8.21}$$

ξ_j is called the *necessity function* of x_j, and takes the form of values from zero to 1. If ξ_j is close to zero, x_j can be removed as unnecessary for evaluation. After it is removed, the model must be built again, fuzzy measure identified, necessity function found, and the process of finding unnecessary attributes repeated. If this is done repeatedly, the evaluation structure can be identified.

REFERENCES

(1) Sugeno, M., "Fuzzy Theory [IV]," *Journal of the Society for Instrument and Control Engineers*, **22**, 6, pp. 50–55 (1983) (in Japanese).
(2) Ishii, K., and Sugeno, M., "A Model of Human Evaluation Process Using Fuzzy Measure," *International Journal of Man-Machine Studies*, **22**, pp. 19–38 (1985).

Chapter 9

DIAGNOSIS

In this chapter we will explain the ambiguity that accompanies fault diagnosis and medical diagnosis and how that ambiguity is handled. There are many proposals and much experimental research about methods for diagnosis that include ambiguity; of these, we will consider only two or three basic methods. First, we will talk about cases in which the relationships between symptoms and causes are expressed in terms of fuzzy relational equations and about the correspondence between the diagnostic process and the solving of inverse problems involving such relationships. Next, we will show a diagnostic method that makes use of the degree of conformity between previously determined abnormal patterns and observations of real problems. Finally, we will touch on approaches by means of knowledge engineering.

9.1 AMBIGUITY IN DIAGNOSIS

When we speak about standard breakdowns or sickness, we mean that either a facility or a human body is in a condition that deviates from the norm. Diagnosis means, in the case of a breakdown, the specifics of content, system, area, and degree; in the case of sickness, it means specifics like the affected part and the degree and name of the sickness. In most cases the diagnosis is tied to measures that bring a return to normal, that is, repair of the breakdown or treatment of the illness.

Especially in the early stages of a breakdown or sickness, there are cases in which we can obtain only ambiguous, often sensory, information such as an intermittent strange sound and the smell of burning, or a slightly high temperature. In general, we can say that the earlier we try to diagnose an abnormality, the more ambiguous the indicators. Whenever we diagnose, it is not uncommon to have circumstances in which we must make some judgment or prediction using this type of ambiguous information. In contrast to plant breakdowns, the objects of medical and nursing diagnosis are people, and there is a need for accurate diagnosis based on the enormous amount of existing knowledge. However, in cases where the difference between normal and abnormal is not clear, such as in early detection of disease and psychological illness, we have to deal with ambiguous information.

Summed up simply, medical diagnosis is the making of judgments about a patient's illness using specialist knowledge, but the observation of symptoms includes results of tests, direct observation of the main complaint, various signs from the patient himself, and the patient's medical history, and, in addition, the circumstances of the diagnosis. For example, if we think about diagnosis during a regular checkup versus before an operation, indications of possible illnesses and need for retesting are important in the former, but in the latter, importance is placed on certainty rather than possibility. Normally an operation is not performed without confirmation of the existence of disease in the affected part.

The words we use concerning symptoms often contain expressions of frequency and probability, such as "in most cases," "every once in a while," "first," and "almost." Especially with expressions like "absolutely," we are inviting untold misunderstandings if we leave out the circumstances and context and take the qualifier as meaning "100%" or "0%." American doctors say, "Never say never," meaning that, even though there are people who think of *never*, which means 0%, as indicating 0–5% depending on how it is received, the more important the circumstances we are in, the more care we must take in choosing an absolutist expression.[1]

In contrast to this kind of linguistic ambiguity, ambiguous circumstances exist in the observation of the patient's condition, such as different laboratories' having differing results in the measurement of the concentration of a substance in a solution, when a value between the normal and abnormal ranges is obtained, or when the data necessary for a standard diagnosis are lacking. In addition it is difficult to make minute classifications of degree and type of pain or unpleasant feeling, and we must resort to ambiguous expressions. In afflictions like psychological barriers, over and above the difficulty of specifying the affected part, there are cases in which the border between normal and abnormal covers a broad range in early stages.

When thinking about the ambiguity that originates in the characteristics discussed above, we must also consider the ambiguity that arises from the participation of people in the evaluation. In general, there is a large dependence on societal values, which are subjective judgments, when we consider things like long-term wear and tear, which is the primary factor in the difficulty of assessing the safety of old buildings and means of transportation.

In this chapter we will consider several diagnostic methods, especially those that deal with those ambiguities, out of all of the above ambiguities, that can be expressed in terms of fuzzy sets; but for detailed explanations of fuzzy relations, their compositions, inference, and compositional rules, please see Chapters 2 and 3.

9.2 DIAGNOSIS USING FUZZY RELATIONS

In this chapter we will consider symptoms, which are indicators of abnormalities, to be the results of a certain cause, and diagnosis as specifying that cause. We will consider some instances in which this assumption is allowable.

Let the set made up of symptomatic items be $Y = \{y_1, y_2, \ldots, y_n\}$. In addition, let r_{ij} express the existence of a connection between x_i and y_j and its depth, and let us consider a matrix of m rows and n columns, $R = \{r_{ij}\}$. In this case, the fuzzy relation is expressed by Equation (9.1), which signifies the operations in Equation (9.2).

$$y = x \circ R \tag{9.1}$$

$$y_j = \bigvee_i \{x_i \wedge r_{ij}\}; \qquad j = 1, 2, \ldots, n \tag{9.2}$$

$\vee$ and $\wedge$ mean max and min respectively.

In general, we find x, given y and relation R, and this corresponds to the diagnostic process. Especially when y_j and r_{ij} are values within the interval $[0, 1]$, this problem is called the *inverse problem* of the fuzzy relational equation. Before getting into the solution, let us think about the meaning of Equation (9.2). For simplicity, let us consider two-valued logic; in other words, $\vee$ and $\wedge$ follow the *and* and *or* of Boolean operations. That is,

When primary i factor occurs, $x_i = 1$.

When symptom j appears, $y_j = 1$.

When primary factor i occurs and symptom j appears, $r_{ij} = 1$.

In all other instances, all are zero.

In this case, we can write

$$y_j = (x_1 \text{ and } r_{ij}) \text{ or } (x_2 \text{ and } r_{2j}) \text{ or } \cdots \text{ or } (x_m \text{ and } r_{mj})$$

$$j = 1, 2, \ldots, n \tag{9.3}$$

for symptom j. When $y_j = 1$, that is, when symptom j appears, at least one of the factors connected with this item occurs, and conversely, when $y_j = 0$ none of the factors connected with this item occur. This can be confirmed directly from Equation (9.3).

Below, we will show how to solve the inverse problem in which we find x_i when all $y_j (j = 1, \ldots, n)$ and r_{ij} are given.

First let us define compositions ω and $\tilde{\omega}$.

Definition 9.1. *For a given $p, q \in [0, 1]$.*

$$p\omega q = \begin{cases} q & \text{if } p > q \\ [q, 1] & \text{if } p = q \\ \phi & \text{if } p < q. \end{cases} \tag{9.4}$$

$$p\tilde{\omega}q = \begin{cases} [0, q] & \text{if } p > q \\ [0, 1] & \text{if } p \leq q. \end{cases} \tag{9.5}$$

In this definition the composition of ω is actually the solution to the inverse problem "p and q are given; find z such that $p \wedge z = q$." From here on ϕ indicates that there is no solution. Using the symbols from this composition, the algorithm for the solution can be divided into three parts:

(1) Find matrices U and $\tilde{U}$.

$$U = \{u_{ij}\} = \{r_{ij}\omega y_j\}. \tag{9.6}$$

$$\tilde{U} = \{\tilde{u}_{ij}\} = \{r_{ij}\tilde{\omega}y_j\}. \tag{9.7}$$

(2) Find matrix $W = \{w_{ij}\}$.

$$w_{ij} = \begin{cases} u_{ij} & \text{for } \exists_1 i \in \{i \,|\, u_{ij} \neq \phi\} \\ \tilde{u}_{ij} & \text{for other } i\text{'s.} \end{cases} \tag{9.8}$$

In the above equation, $\exists_1 i$ means selecting only one element from row i such that $\mu_{ij} \neq \phi$ for each column j. In general, since there are many possible ways to make this selection, we can find many types of matrix W. In order to distinguish these, we use the superscript k, and write $W^k = \{w_{ij}^k\}$.

(3) For each k, we calculate

$$x_1^k = \bigcap_j w_{ij}^k; \qquad i = 1, 2, \ldots, n, \tag{9.9}$$

which gives us the solution.

When not even one solution for W in Equation (9.8) of (2) exists, it means that there is no solution that satisfies the original equation.

The above algorithm can easily be extended to cover cases in which the elements of R and y are given by intervals within $[0, 1]$.

Example 9.1. Let R and y be given as follows:

$$R = \begin{pmatrix} 1 & 0 \\ 1 & 0.5 \\ 0.3 & 1 \end{pmatrix} \qquad y = (1 \quad 0.5).$$

(1) Find U and $\tilde{U}$.

$$U = \begin{pmatrix} 1 & \phi \\ 1 & [0.5, 1] \\ \phi & 0.5 \end{pmatrix} \qquad \tilde{U} = \begin{pmatrix} - & - \\ - & - \\ - & [0, 0.5] \end{pmatrix}.$$

(2) Find W.

$$W^1 = \begin{pmatrix} 1 & - \\ - & [0.5, 1] \\ - & [0.5, 1] \end{pmatrix} \qquad W^2 = \begin{pmatrix} 1 & - \\ - & - \\ - & 0.5 \end{pmatrix}$$

$$W^3 = \begin{pmatrix} - & - \\ 1 & [0.5, 1] \\ - & [0, 0.5] \end{pmatrix} \qquad W^4 = \begin{pmatrix} - & - \\ 1 & - \\ - & 0.5 \end{pmatrix}.$$

(3) Find x.

$$x^1 = (1, [0.5, 1], [0, 0.5]),$$

$$x^2 = (1, [0, 1], 0.5),$$

$$x^3 = ([0, 1], 1, [0, 0.5]),$$

$$x^4 = ([0, 1], 1, 0.5).$$

We can confirm that these four solutions all satisfy the original fuzzy relational equation by actually calculating $y = x^k \circ R(k = 1, 2, 3, 4)$. In the case of diagnosis, out of these four solutions, x_3 is excluded, but x_1 and x_2 are judged to be occurring.

9.3   DIAGNOSIS USING SYMPTOM PATTERNS AND DEGREES OF CONFORMITY

In the last section we discussed cases in which it is acceptable to think of there being a cause of disease or breakdown, with the symptoms appearing as a result, but in this one we will consider cases in which a specific group of symptoms is indicated and the disease named. In this case diagnosis means investigating the degree of conformity of previously specified symptom patterns and actual observations.[2]

We let $\{A_j, j = 1, 2, \ldots, n\}$ be a group of names of attributes connected with symptoms. These include temperature, concentration, etc. We let A_j take the form of linguistic expressions such as "very high," and we express the support set of this kind of fuzzy set by U_j. In addition, we let the fuzzy set that is characterized by membership function $\mu_{Rj}: U_j \to [0, 1]$ be expressed by R_j. Fuzzy proposition P_j, which is something like "high temperature" can be written as follows:

$$P_j = \text{``}A_j \text{ is } R_j\text{''}; \qquad j = 1, \ldots, n. \tag{9.10}$$

In addition, compositional proposition

$$S = \text{``}P_1 \text{ and } P_2 \text{ and } \ldots \text{ and } P_n\text{''} \tag{9.11}$$

can be written simply as

$$S = \text{``}A \text{ is } R\text{''}. \tag{9.12}$$

$A = (A_1, \ldots, A_n)$, and R is an n-dimensional direct product of fuzzy sets from R_1 to R_n. Now, let us consider Equation (9.12) and a symptom pattern that corresponds to one particular disease, and on the other side, let the actual observational data for the object be given as "A_j is Q_j" for each A_j. As noted before, we write

$$O = \text{``}A \text{ is } Q\text{''}. \tag{9.13}$$

Q is a direct product of fuzzy sets from Q_1 to Q_n.

In order to diagnose whether the disease in question is occurring or not, we consider the coexistence of two compositional propositions S and O; that is, we consider the degree of conformity of R and Q. The conformity of two fuzzy sets is expressed by γ, and we solve the following equation:

$$\gamma = \sup_{u \in U} (R \cap Q). \qquad (9.14)$$

$U = U_1 \times U_2 \times \cdots \times U_n$, that is, a standard direct product.

Here we must be careful about the following. We define the direct product of fuzzy sets as the min, and γ_j is determined by

$$\gamma_j = \sup_{u_j \in U_j} (R_j \cap Q_j). \qquad (9.15)$$

Equation (9.14) can be expressed as follows:

$$\gamma = \min_j \{\gamma_j\}. \qquad (9.16)$$

In other words, γ is the minimum value out of the conformities of R_j and $Q_j (j = 1, 2, \ldots, n)$, and if the conformity for even one of the attributes is low, it has a direct effect. This is because the direct products of fuzzy sets and connections with "and" are determined by the min, so we can consider a slightly looser definition. We could consider another method of determination, but this section is aimed at an understanding of the basic way of thinking, and a number of devices are necessary to expand on this point. Furthermore, even if we improve the "and" operation, we cannot completely express the connections between symptoms, when they take the form of a consideration of concrete details, using the above method. As a simplified example, let us consider the case of the common cold. The uncertainty of a cold is greater in the case of both "high temperature" and "much coughing" than when either symptom appears by itself. Here, we could consider using a method that starts by considering these two symptoms as being separate things. If we consider a generalization of this example for n number of attributes, we have $(2^n - 1)$ number of symptoms. Naturally, the degree to which a sickness exists is increased by a combination of important symptomatic items. We let this kind of knowledge make up fuzzy set H, which is expressed by the following membership function:

$$\mu_H: 2^K \to [0, 1]. \qquad (9.17)$$

$K = \{A_1, A_2, \ldots, A_n\}.$

Fuzzy set H, for which we let the power set of the sets of attributes be the support set, is the knowledge about medical diagnosis itself. If we can specify this kind of knowledge in some form beforehand, we can construct a diagnostic method that makes overall judgments from it and from fuzzy data Q, the observed condition of the patient.[3]

9.4 APPLICATIONS OF KNOWLEDGE ENGINEERING IN DIAGNOSIS

As far as the use of knowledge engineering in medical diagnosis goes, since Shortliffe[4] publicized his system called MYCIN, which had as its object diagnosis and treatment of symptoms of bacterial infection, in 1976, more than 20 systems have been proposed, of which there are outstanding explanations.[5]

There are two large parts to what are called knowledge engineering systems. That is, there is the collection of knowledge expressed in the form "IF A THEN B," and the inference mechanism that corresponds to the use of that knowledge. MYCIN has been put to use for medical diagnosis and treatment, but in its development it has been tied to the system called TEIRESIAS by R. Davis, and furthermore, in 1980 W. van Melle constructed the system called EMYCIN, the structure of which goes beyond medical diagnosis can be used very widely. If the knowledge of an expert in any field—for instance, that of business or nursing diagnosis—can be expressed in the form of "IF A THEN B," it can be used. Since this kind of system puts the knowledge of experts or the experience of specialists into the memory of a computer and uses it, it is called an *expert system*, and many researchers are heaping attention on such systems.[7]

There are cases in which the A and B of "IF A THEN B" are fuzzy propositions, and those in which the veracity of "IF A THEN B" itself is not complete. K. P. Adlassing *et al.* developed an expert system called CADIAG (Computer Assisted Medical Diagnosis) in which diagnoses are made with the ambiguity left in.[8] We can consider the following four items in this system:

S: symptoms, indicators and test results

D: name of disease or diagnosis

SC: combinations of symptoms

IC: internal combinations.

We let ϕ mean *unknown* or *undetermined*, and T is determined by the following equation:

$$T = [0, 1] \cup \phi. \tag{9.18}$$

Each symptom S_i is characterized by μ_{S_i}, which has a value in this T, and its size is understood to be the degree to which the symptom appears. However, unlike standard membership functions, it takes the value ϕ; and the intersection, union, and complement of the fuzzy set are defined as follows:

$$X_1 \wedge X_2 = \begin{cases} \min\{x_1, x_2\} & \text{if } X_1 \in [0,1] \text{ and } X_2 \in [0,1] \\ \phi & \text{if } X_1 = \phi \text{ and/or } X_2 = \phi \end{cases}$$

$$X_1 \vee X_2$$
$$= \begin{cases} \max\{x_1, x_2\} & \text{if } X_1 \in [0,1] \text{ and } X_2 \in [0,1] \\ X_1 & \text{if } X_1 \in [0,1] \text{ and } X_2 = \phi \\ X_2 & \text{if } X_1 = \phi \text{ and } X_2 \in [0,1] \\ \phi & \text{if } X_1 = \phi \text{ and } X_2 = \phi \end{cases} \tag{9.19}$$

$$\bar{X}_1 = \begin{cases} 1 - X_1 & \text{if } X_1 \in [0,1] \\ \phi & \text{if } X_1 = \phi \end{cases}$$

In addition, we can consider the following fuzzy relations between items: symptom–disease (SD), symptom combination–disease ((SC)D), symptom–symptom (SS) and disease–disease(DD). The depth of the relation between each element is given from two sides; frequency and strength of confirmation. For example, in order to make SD, we write the knowledge that relates to symptom–disease in the following form:

$$\text{IF } A \text{ THEN } B \text{ WITH } (O, C) \tag{9.20}$$

A is the term related to the symptom and B the term related to diagnosis. The linguistic expressions found in Table 9.1 can be used for O and C, but finally the various fuzzy relations are constructed using the representative numbers. If the data concerning the patient are given, all we have to do is perform the composition of the fuzzy relations and fuzzy data. In other words, using the compositional rules from Zadeh's inference,[9] confirmation or exclusion of the sickness can be carried out.

As for applications in fuzzy set diagnosis, an important point is how to handle the ambiguity that arises within the diagnostic process. We must look to the results of specialized research in each field for the necessary knowledge and logic for diagnosis. However, research into the acceptability of the kind of ambiguity discussed in this chapter and into suitable methods of expression

Table 9.1. Linguistic Values, Intervals and Representative
Values Expressing Frequency and Strength of Confirmation

Linguistic Value (Strength of Confirmation $m(\ )$)		Interval	Representative Value
always	(always)	$[1, 1]$	1
almost always	(almost always)	$[0.98, 0.99]$	0.99
very often	(very strong)	$[0.83, 0.97]$	0.9
often	(strong)	$[0.68, 0.82]$	0.75
sometimes	(medium)	$[0.33, 0.67]$	0.5
seldom	(weak)	$[0.18, 0.32]$	0.25
very seldom	(very weak)	$[0.03, 0.17]$	0.1
almost never	(almost never)	$[0.01, 0.02]$	0.01
never	(never)	$[0, 0]$	0
unknown	(unknown)	ϕ	ϕ

for that ambiguity will become more and more important from now on. There
are still few knowledge engineering systems that are actually on the scene. It is
necessary to develop fuzzy expert systems to process ambiguity appropriately
and simultaneously with the construction of systems that approach the level
of ability of specialist doctors.

REFERENCES

(1) Tsukamoto, Y., "Fuzzy Diagnosis," *Operations Research*, **26**, pp. 698–704 (1981)
    (in Japanese).
(2) Sanchez, E., "Medical Diagnosis and Composite Fuzzy Relations," *in Advances
    in Fuzzy Set Theory and Application*, Gupta, M. M., *et al.*, eds., North-Holland,
    Amsterdam (1979) pp. 437–444.
(3) Tsukamoto, Y., and Clouaire, R. M., "A Method for Diagnosis under Uncertainty
    Using Fuzzy Integrals," *Proc. of 2nd Conference, Medical Information Association*,
    pp. 155–158 (1982) (in Japanese).
(4) Shortliffe, E. H., *Computer Based Medical Consultations MYCIN*, Elsevier (1976).
(5) Koyama, T., and Kaibara, N., "Applications of Knowledge Engineering to Medical

Diagnosis," *Journal of the Society of Instrument and Control Engineers*, **23**, 4, pp. 353–360 (1984) (in Japanese).

(6) Davis, R., *Knowledge-Based Systems in Artificial Intelligence*, Mcgraw-Hill, New York (1982).

(7) Expert System Special Number, *Information Sciences*, **37**, 1–3, pp. 1–292 (1985).

(8) Adlassnig, K. P., Kolarz, G., and Scheithauer, W., "Present State of the Medical Expert System CADIAG-2, *Method of Information in Medicine*, **24**, 1, pp. 13–20 (1985).

(9) Zadeh, L. A., "Linguistic Variables, Approximate Reasoning and Dispositions," *Medical Information*, **8**, 3, pp. 173–186 (1983).

Chapter 10

CONTROL

When the ideas of fuzzy logic are applied to control, it is generally called *fuzzy control*. Fuzzy control was the first application of fuzzy theory to get attention, and it is a field in which research has forged ahead. Control of cement kilns, electric trains, water purification plants, and other facilities is actually being carried out. Here we will center on the tools necessary to carry out fuzzy control and give an outline of the thinking behind it.

10.1 THE FORM OF FUZZY CONTROL RULES AND INFERENCE METHODS

Fuzzy control describes the algorithm for process control as a fuzzy relation between information about the condition of the process to be controlled, x and y, and the input for the process (amount of work), z. The control algorithm is given in "if–then" expressions, such as

If x is small and y is big, then z is medium.
If x is big and y is medium, then z is big.

159

These are called *fuzzy control rules*. The "if" clause of the rules is called the *antecedent* and the "then" clause the *consequent*. In general, variables x and y are called the *input* and z the *output*. "Small" and "big" are fuzzy values for x and y (sometimes called fuzzy variables), and they are expressed by fuzzy sets.

Fuzzy controllers are constructed of groups of these fuzzy control rules, and when an actual input is given, the output is calculated by means of *fuzzy inference*. Fuzzy inference is based on fuzzy logic, but in consideration of the time for calculation, simple methods are used. The inference for fuzzy control is different from standard fuzzy inference in that the propositions (the actual input for the fuzzy controller) are commonly standard numerical values, not fuzzy values. The major difference between the methods used in areas such as production rules for knowledge engineering and fuzzy control is that the latter permits fuzzy expressions (single stage inference) whereas the former is almost always multistage.

In the following, for simplicity, we will use the notation $A(x)$ for the membership function of fuzzy set A. The forms taken by fuzzy control rules are classified by three points—the form of the antecedents and the consequents, the form of fuzzy variables, and the inference method—but these are not necessarily independent. We will discuss these based on the following three inference methods.

10.1.1   Inference Method 1

There are two types of fuzzy variables, continuous and discrete. Continuous variables are shown in Fig. 10.1(a) and (b). (a) shows a bell type and (b) a triangular type, and both of them specify a fuzzy variable with two parameters. NB, ZO, PS, etc. indicate meanings like Negative Big, Zero, and Positive Small. These express fuzzy subsets (or fuzzy numbers) of the interval $[-1, 1]$. In the case of fuzzy control, the domain of input and output variables that can have either positive or negative values is commonly standardized to $[-1, 1]$, and that for those that can have only positive values to $[0, 1]$. Be-

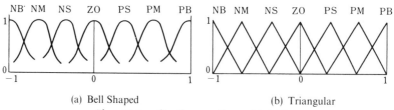

(a) Bell Shaped   (b) Triangular

Fig. 10.1.   Continuous Fuzzy Variables

Table 10.1.   Discrete Fuzzy Variables

	-6	-5	-4	-3	-2	-1	0	$+1$	$+2$	$+3$	$+4$	$+5$	$+6$
PB	0	0	0	0	0	0	0	0	0	0	3	7	10
PM	0	0	0	0	0	0	0	0	3	7	10	7	3
PS	0	0	0	0	0	0	3	7	10	7	3	0	0
ZO	0	0	0	0	3	7	10	7	3	0	0	0	0
NS	0	0	3	7	10	7	3	0	0	0	0	0	0
NM	3	7	10	7	3	0	0	0	0	0	0	0	0
NB	10	7	3	0	0	0	0	0	0	0	0	0	0

cause of this, we can use similar fuzzy variables for all variables. Table 10.1 is an example of discrete fuzzy variables. However, the grade is expressed by integers from zero to 10. The domain of the input and output variables is discrete in the range of integers from -6 to 6, and a single fuzzy variable is specified by three parameters.

In fuzzy inference method 1, it is common to have five to seven fuzzy variables. However, adjustment of parameters is not really necessary, and standard variables can be used.

A two-input, single-output instance of this inference method comes out as follows. We set up the following two control rules:

$$\text{If } x_1 \text{ is } A_{11}, \qquad x_2 \text{ is } A_{12}, \qquad \text{then } y \text{ is } B_1. \tag{10.1}$$

$$\text{If } x_1 \text{ is } A_{21}, \qquad x_2 \text{ is } A_{22}, \qquad \text{then } y \text{ is } B_2. \tag{10.2}$$

An example of this kind of rule is single-input, single-output process set-point control, and we can consider the case in which e and Δe (the change in e during one sampling) are chosen for the controller input, and Δu (the change in control) for the output.

Now let the input be $x_1 = x_1^\circ$ and $x_2 = x_2^\circ$. First we find the compatibility for each of the antecedent conditions of the rules and the input. In general, we let the compatibility for antecedent "x is A" be $A(x^\circ)$, that is, the membership function of x° for fuzzy set A. Here, the antecedent is two-dimensional, so we let compatibility be

$$\omega_i = A_{i1}(x_1^\circ) * A_{i2}(x_2^\circ); \qquad i = 1, 2. \tag{10.3}$$

i is the number of the rule, and $*$ is multiplication.

Next, we let the results of inference for the ith rule be

$$y \text{ is } \omega_i B_i, \qquad \text{but } \omega_i B_i(y) = \omega_i \times B_i(y). \tag{10.4}$$

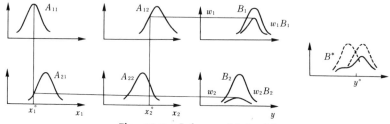

Fig. 10.2. Inference Method 1

There are cases where a min operation, as in

$$\omega_i B_i(y) = \omega_i \wedge B_i(y), \tag{10.5}$$

replaces the multiplication.

The complete inference result y° is constructed from $\omega_1 B_1$ and $\omega_2 B_2$,

$$B^* = \omega_1 B_1 \cup \omega_2 B_2, \tag{10.6}$$

and found in terms of the central axis of the membership function of B^*. In other words, we get

$$y^\circ = \int B^*(y) y \, dy \Big/ \int B^*(y) \, dy. \tag{10.7}$$

Figure 10.2 shows this inference method.

The above inference process is made up of the following three steps. These are essentially the same for all inference methods.

(1) Calculate the compatibility for the input and antecedents of the rules.
(2) Find the inference results for each rule.
(3) Find the complete inference result as a weighted mean of the inference results for each rule with respect to compatibilities.

10.1.2   Inference Method 2

This method is especially good for fuzzy variables that have monotonic membership functions, such as those shown in Fig. 10.3. As can be seen from the figure, there are only two types of variables, Positive and Negative, but we let arctan(x) be used for the membership function; and there are changes in slope.

As an example, let us consider the following two rules.

$$\text{If } x_1 \text{ is } N, \qquad x_2 \text{ is } P, \qquad \text{then } y \text{ is } N. \tag{10.8}$$

$$\text{If } x_1 \text{ is } P, \qquad x_2 \text{ is } N, \qquad \text{then } y \text{ is } P. \tag{10.9}$$

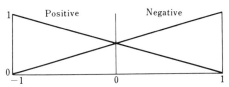

Fig. 10.3. Monotonic Fuzzy Variables

The compatibility of the antecedents for inputs x_1° and x_2° is found as in the previous method, and we let them be ω_1 and ω_2. Inference results y_1 and y_2 (nonfuzzy values) for each rule are found using the following relational equations:

$$\omega_1 = N(y_1), \qquad \omega_2 = P(y_2). \tag{10.10}$$

The overall inference result is given by

$$y^\circ = \frac{\omega_1 y_1 + \omega_2 y_2}{\omega_1 + \omega_2}, \tag{10.11}$$

by taking the weighted mean of y_1 and y_2. The inference process is shown in Fig. 10.4.

It is generally said that fewer rules are necessary for this method than for the first one, and that it is appropriate when there are many input variables. However, since there are few fuzzy variables, it is not a very appropriate method for putting the knowledge of experts, which takes linguistic form, into logical form.

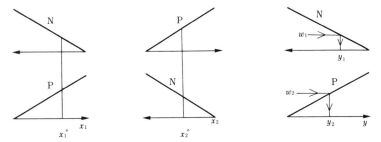

Fig. 10.4. Inference Method 2

10.1.3 Inference Method 3

The antecedents used in this method are made up of fuzzy propositions, and the consequents are the standard relational equations of inputs and outputs. This was conceived for fuzzy process modeling rather than for fuzzy control.

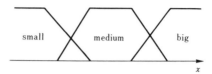

Fig. 10.5. Trapezoidal Fuzzy Variable

The fuzzy variables used in the antecedents are, as shown in Fig. 10.5, ones that have trapezoidal membership functions constructed from straight lines. Let us consider the following two rules.

$$\text{If } x_1 \text{ is } A_{11}, \qquad x_2 \text{ is } A_{12}, \qquad \text{then } y = f_1(x_1, x_2). \qquad (10.12)$$

$$\text{If } x_1 \text{ is } A_{21}, \qquad x_2 \text{ is } A_{22}, \qquad \text{then } y = f_2(x_1, x_2). \qquad (10.13)$$

If we let the compatibility of the antecedents for x_1° and x_2° be ω_1 and ω_2, the inference results for each of the rules are calculated directly from the equations written in the consequents. The complete inference result is found using the following equation, as in inference method 2:

$$y^\circ = \frac{\omega_1 f_1(x_1^\circ, x_2^\circ) + \omega_2 f_2(x_1^\circ, x_2^\circ)}{\omega_1 + \omega_2}. \qquad (10.14)$$

Here, f is usually a linear relational equation. If the number of rules is 1, the antecedent parts are no longer necessary, and only the consequent part remains; so the result is the same as having a linear expression. If there is more than one rule, the input interval is partitioned into subspaces and a linear input/output relation is found for each subspace; that group gives us something very similar to the global nonlinear input/output relation. This method is not appropriate for linguistic expressions, but exceeds the others in descriptive capability. The rules used in inference method 1 do not go beyond description of quantitative relations.

For example, there is no difference in the structure of the relation

$$\text{if } x_1 = PB, \qquad x_2 = PS, \qquad \text{then } y = NM$$

and

$$\text{if } x_1 = 10, \qquad x_2 = 2, \qquad \text{then } y = -6.$$

In other words, they are numerical tables and only quantify amount. In contrast, the rules that arise using the conditions (subspaces) specified in the antecedents in the third form are written directly in the consequents.

As we have already mentioned, the antecedents in the rules of all three forms are most easily understood when interpreted as being ambiguous partitionings of the input spaces, that is, specifiers of fuzzy subspaces, rather

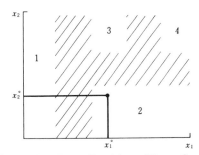

Fig. 10.6. Fuzzy Partition of Input Space

than descriptions of conditions. Fig. 10.6 shows the situation when a two-dimensional input space is partitioned four ways. The oblique lines indicate that the boundaries are ambiguous. As is quickly understood, there only as many rules as there are partitions. In addition, the compatibility of the antecedents is nothing other than the degree to which the input (x_1°, x_2°) belongs to the fuzzy subspace specified by the antecedent, that is, the membership value.

10.2 PLANNING OF FUZZY CONTROLLERS

The problem when fuzzy control is to be used in an actual process is the design of the controller. To design a controller means to determine the form of the control rules and to write them out concretely, and the problem can be divided into two parts, determination of the antecedents and determination of the consequents. As far as the antecedents are concerned, three things have to be determined. First, the input information for x_1, x_2, etc., which have to be used in the antecedents, is selected; next is the determination of the conditions, that is, the fuzzy partitions of the input; and third is the determination of the parameters for the fuzzy variables. As far as the consequent is concerned, the output is generally the control input for the process, which is determined of its own accord. The only remaining problem is the fuzzy parameters. Therefore, determination of the consequents is not difficult, and the problem is wholly the determination of the antecedents. Generally speaking there are three design methods for this.

10.2.1 Expert Experience and Knowledge

This is the expert systems approach. One can say that fuzzy control is actually the first real example of an expert system. The experience of skilled operators and the knowledge of control engineers is expressed qualitatively in words, and if these are put into logical forms as fuzzy control rules, a controller can be planned.

Of the problems in determining the antecedents of the rules, the information input to the fuzzy controller for inference of the control input becomes clear of its own accord and is no problem. The main problem is the fuzzy partitions of the input space, which must basically be determined through interviews with operators and by using the instincts of control engineers. If we use inference method 1, which is appropriate for this type of design, we need not worry much about the parameters for the fuzzy variables.

10.2.2   Operator Models

The operation of complicated processes is performed skillfully by experts, but it is not always easy to put their know-how into logical form. For example, experts may not be able to express their work in words. Even if they can, there are instances when their words express things incompletely. As can be seen with driving a car, when an expert learns an operation with his or her hands and feet, it is next to impossible to express the skills in words. There are also times when we cannot obtain the cooperation of onsite operators.

An effective design method in this type of case is a model of the functions carried out by the operator. This means making a model of the input/output relation between the information used by the operator and his functional output. If an "if–then" form like that of control rules is chosen for the model, it can be used as is, as a fuzzy controller. Identification of the model uses the operator's actual input and output.

In this case too, the input information presents little problem, because we practically know what information the operator is using. The problem is the fuzzy partitions, for which we have to solve a so-called structure identification problem using the input and output data. In this method we must identify the fuzzy parameters, but this problem in itself is no different from a standard identification problem.

10.2.3   Fuzzy Models of Processes

The two above methods are ones in which an expert model is constructed and forms a fuzzy controller, but these models cannot surpass the experts. When the object is a process without experts or human operators, a better method is based on a fuzzy model of the process for the design of a controller aimed at high quality control.

Here a fuzzy model means describing the features of the process using an "if–then" form that is the same as that of fuzzy control rules. One "if–then" expression is called a *process action* or *process rule*, the fuzzy model being a number of these process rules grouped together. The concepts of a fuzzy control system are shown in Fig. 10.7. First, let us consider the identification of fuzzy models. Using inference method 3, in which the consequent is a description of the input/output relation for the process, gives us a form of the model that is convenient for cases in which a high-order multivariate system is the object. Identification is divided into that of the antecedent and the consequent, but the identification of the conclusion is essentially the same as identification of a linear model. The consequent for each process rule corresponds to the linear models used up to now, so there is an increase in the amount of calculations. As for the structure of the antecedent, we must consider the following two things:

(1) selection of input variables, i.e., which input variables out of all the process inputs should go into the antecedents;
(2) fuzzy partitions, a problem similar to the above problem of partitioning the input space.

There is no established method for identifying these, and (2) is an especially difficult problem.

There are two possible ways of thinking to design of fuzzy controllers from fuzzy models; both use sets of process rules that describe partial actions of the process.

(1) process rule × control object = control rule.
(2) process rule + control rule = ideal process action.

(1) is the method of finding the control rule that, for example, minimizes a particular evaluation function, and (2) is a method that compensates for process rules (actions) by means of control rules, and seeks to bring about the best possible process action. The idea common to both is making the characteristics of the fuzzy control system expressions in Fig. 10.7 work well by

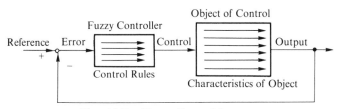

Fig. 10.7. Fuzzy Control System

finding one control rule that corresponds to each process rule. Since one rule expresses a partial action, the corresponding control rule is partial in nature (in a fuzzy subspace) and made conditional by the "if."

10.3   FEATURES OF FUZZY CONTROL

Fuzzy control has the following three features:

(1) logical control;
(2) parallel (dispersed) control;
(3) linguistic control.

It goes without saying that the "logical" of (1) means the free expression of control algorithms using the "if–then" form. The "if" clause, in particular, can describe a wide variety of conditions using logical combinations with "or" and "and." The arranged control of (2) means that general control policies can be made to work in a dispersed manner by means of a control rules. This differs qualitatively from the single-equation methods used up to now, and we might say that the coexistence of controls with differing logics is possible. (3) means that it is possible to use ambiguous linquistic variables, especially in the antecedents of the rules. Language is easy for people to understand qualitatively, and control using dialogue with operators is possible. In addition, through effective use of features (1)–(3), it will be possible to use the experience-trained "eyes" of operators on observation processes as external inputs and things like process conditions as effective information for control. In addition, the unusual procedures that always accompany the operations of a real process can be brought into the control algorithm.

REFERENCES

(1) Yamazaki, Sugeno, "Fuzzy Control," *System and Control*, **28**, pp. 18–22 (1984) (in Japanese).
(2) Sugeno, M., "An Introductory Survey on Fuzzy Control," *Information Sciences*, **36**, pp. 59–83 (1985).
(3) Sugeno, M., ed., *Industrial Applications of Fuzzy Control*, North-Holland (1985).
(4) Miyamoto, "Fuzzy Control and Its Applications," *Systems and Control*, **25**, 5, pp. 458–465 (1986).

Chapter 11

HUMAN ACTIVITIES

Most plant and transport breakdowns happen because of some form of human error. We need to study human reliability in order to increase safety. Up to now, human error and mechanical breakdowns have been approached in a similar manner and carried out using probability expressions.[1] However, human beings are different from machines and are influenced by an extremely large number of factors; their reactions are widely varied except in cases of simple repetitive work, so it is impossible to express human reliability in terms of probability. In addition, one person can accomplish several jobs in parallel, and can cooperate with others to perform complicated tasks. There are many aspects and functions such as learning, judgment, and reasoning that we cannot discuss on the same level as mechanics.

In this chapter we will consider structural models and arrangements of fuzzy sets as human reliability models instead of probability. Both of these are qualitative models and do not produce the kinds of numerical values that probability does, but they are appropriate for expressing the essentials of complicated problems. When dealing with a problem with complicated ambiguity, as when people are the object, the proper approach is first to get a general understanding of the essential aspects, and then quantitatively to analyze minute points.

169

11.1   HUMAN RELIABILITY MODELS

In a previous publication, the authors discussed a human reliability model for the case of several people cooperating to accomplish one task.[2] In addition, numerous experiments were conducted on the converse, in which one person performs several tasks in parallel, and a structural model was proposed.[3,4] This model is made up of a physiological part and a psychological part. The former mainly shows the effect of the amount of work on human reliability, while the latter expresses the effects of the concentration of care or tension. Naturally, structural models must be objective and general, even though they are normally difficult to prove. In our study, objectivity was accomplished by devising experiments. In order to make a structural model more precise, modeling must be done for the characteristics of each of the structural elements, but as mentioned before, there are so many influential factors that mathematical models cannot be used. Therefore, we let the characteristics be expressed in rough terms by fuzzy sets, and we tested the identification of their membership functions by experiments. As a result we obtained a human reliability model that can also be used in cases where many people perform multiple tasks.

The experiments performed were as follows. Three random numbers were shown on a CRT, and the subjects responded by pressing the key for the last digit of their sum. Both wrong answers and failures to answer within a certain time limit were counted as failures. The time limit corresponded to workload. The task was repeated 100 times, and the percentage of correct answers was defined as the reliability of the subject. In addition, from the results of previous experiments, the factors related to human reliability were known to be workload, ability, ability distribution among multiple tasks, and psychological stress. Five interrelations among factors can be brought up: (1) workload, ability, and reliability (a three-way relation) (2) workload and stress, (3) stress and ability, (4) ability and distribution, and (5) environment and stress.

Since item (1) is the most ambiguous, we let it be expressed by fuzzy sets. Thus, the ambiguous part of the reliability is brought out in the membership function, and the other parts can be logically streamlined. If we now let the workload be the universe of discourse and think of the reliability corresponding to it as the value of the membership function, the characteristics of humans are very similar to those in Fig. 11.1. Changes in ability can be expressed as the horizontal movement of the membership function. However, an extremely large number of experiments using a single subject is necessary in order to find the membership function itself. In order to avoid this obstacle, the number

Fig. 11.1. Membership Function for Human Reliability

of successful attempts within a given work time is examined, and the relation between the inverse of that time and the success ratio found.[5] These are the bold lines in Fig. 11.2. Since the inverse of the work time can be thought of as the pressure felt by the subject, pressure is a proportion of the workload, and Fig. 11.2 itself is a model of factor (1). In addition, changes in the time limit produce horizontal movements, and this point is fortunate in that it resembles Fig. 11.1.

The relation between workload and stress of factor (2) is normally viewed as a proportional relation. The problem is factor (3), the stress/ability relation. In the psychological field it is said that the proper amount of stress maximizes human ability. However, there are individual variations in ability itself, so here we use the integrated value of reliability shown in Fig. 11.2 to define the person's ability. If we investigate the degree to which ability changes in relation to the inverse of the time limit (workload) for all subjects, we get the

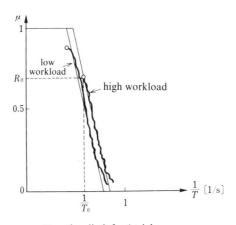

T_0: time limit for 1 trial
R_0: reliability at T_0
μ: rate of success in time T

Fig. 11.2. Example of Membership Function Identification

average response shown in Fig. 11.3, which agrees with results from psychological studies. From the above results, we can see that a structural model for one person with one job would be something like Fig. 11.4.

Next, in the case of one person with two jobs, we add the problem of how the subject distributes his total ability to the two jobs. This varies with the type of work, the workload, and the personality type of the subject.[3] The curves for jobs I and II of Fig. 11.5 show an example for two jobs of the same type and with even workloads. If the same subject performs a single job with double the workload, we get the line on the right. This agrees with the sum of jobs I and II (dotted lines). From the above results, we can see that the overall ability for two jobs equals the sum of the abilities for each job, and that for equal workloads, the distribution of ability is equal.

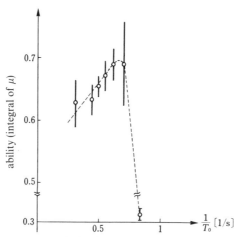

Fig. 11.3. Relation between Workload and Ability

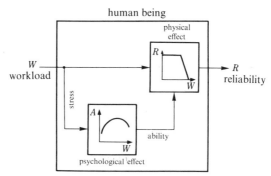

Fig. 11.4. 1 Person/1 Job Reliability Model

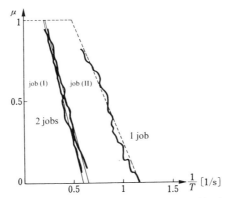

Fig. 11.5. 1 Person/2 Job Ability Distribution

If the total workload remains fixed, but the workloads for each job are different, the distribution of ability differs with the personality type of the subject. Table 11.1 shows a three-way division of subjects according to type.

Bringing together these results, we get a one person/two jobs reliability model like that in Fig. 11.6. Using this model, we can investigate a two person three jobs reliability problem. Of the three jobs, each person is in charge of one, and the one that is left is the one on which the two people cooperate. In

Table 11.1.   Human Personality and Ability Distribution

Personality	Ability Distribution	Percentage of People
perfectionist	concentrate ability on hard work	12%
utilitarian	concentrate ability on easy work	35%
balance type	work equally regardless of workload	53%

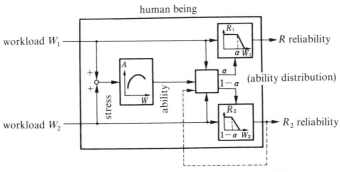

Fig. 11.6. 1 Person/2 Job Reliability Model

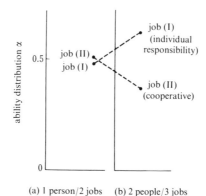

(a) 1 person/2 jobs   (b) 2 people/3 jobs

Fig. 11.7.   2 Person/3 Job Individual Reliability

other words, this a system has parallel redundancy, and if one of the two gives a correct answer, the operation is a success. The workloads for the three jobs are equal. Figure 11.7 shows the results of the experiment for one of the subjects. Figure 11.7(a) shows the results in the case of one person and two jobs, and it is just about the same as Fig. 11.5. Fig. 11.7(b) is the case of two people and three jobs, and if clearly shows that the subject neglected the common work and distributed the greater part of his ability to the job he was responsible for. The subject appeared to adjust the distributions of ability during work when he saw the results. The dotted lines in Fig. 11.6 represent this feedback.

As in the above, the model for Fig. 11.6 explains the experimental results well, and we think it can be used for human-machine problems.

11.2   DATA ENTRY SYSTEMS

One of the important problems in man-machine systems MMS is the exchange of information between humans and machines (computers). Here we will investigate the case in which people input information and propose a new data entry system.

The following two problems arise when people input information, in question-and-answer or any other form, into a computer.

(1) There is not enough information on the people side, and the inputting of ambiguous information cannot be avoided.

(2) When, in spite of the fact that there is only ambiguous information, a selection must be made from a number (usually 3–7) of choices (conditions),

and one does not have enough confidence to decide whether to give the same answer when a question is repeated. From a different point of view, we can ask what is the best number of choices (hereafter, number of conditions) for obtaining detailed information from people. In addition, there generally is the illusion that the larger the number of conditions, the more detailed the information, but when the information is ambiguous, the danger arises of a lack of reproducibility and reliability.

We can see that problem (1) can be solved by introducing the membership concept from fuzzy theory when we consider ambiguity along with a {yes, no}, two-choice arrangement. However, in a really ambiguous case, it is impossible to specify a single fixed membership value from the interval $[0, 1]$. Taking this point into consideration, several alternatives have been investigated: solution policies, such as the method of using linguistic variables (which uses expressions such as "seems to be just about" and "don't know" (D. K.), rather than values from the interval $[0, 1]$); the method of using type II fuzzy sets (or their generalization type N fuzzy sets), which specify one function on the interval $[0, 1]$; or the method of using the concept of probabilistic sets, in which randomness is added to fuzziness. All of these were proposed in the 1970s and have been used in various fields with good results, but for MMS data entry, we will discuss the method of using the concept of probabilistic sets, bearing in mind that this method is easy for people to become familiar with and for computers to process.

As one extension form for fuzzy sets; probabilistic sets have several methods of expression, such as probabilistic expressions and extended fuzzy expressions.[6] Here we will use extended fuzzy expressions. These satisfy three mathematical conditions and are defined by a set of a countably infinite number of functions called monitors.[6]

$$\{M, V = M^2, M^3, M^4, \ldots\}. \tag{11.1}$$

Theoretically, monitors are a countable number of function sets, but the ones that have important information are lower monitors. M is a membership function, and V is called a vagueness function. It is known that these two express most of the important information. In addition they are functions with values from $[0, 1]$, and M corresponds to Zadeh's membership concept. Also, vagueness function V is something that suggests the degree of vagueness given for the value of M in values on $[0, 1]$. (For example, when we have the condition of being absolutely confident in a characteristic's being 80% satisfied for an object under evaluation, we get $M = 0.8$, $V = 0$; but when we have no confidence in the reproducibility and make an evaluation of complete vagueness, we get $M = 0.8$, $V = 1.0$. If we make the evaluation about half-confident, we get $M = 0.8$, $V = 0.5$.)[7]

From the above point of view, we can see that with the data entry system previously used for MMS, only the membership function was input, causing the inaccuracy of information mentioned in the problem section. Therefore, we propose a new data entry system in which the person answering enters two pieces of information; the answer and the degree of confidence he has in giving that answer (vagueness). Figure 11.8 gives a concrete example of the use of this system. (In this example the answer is "a rough feeling" of having "not much" of an appetite.)

Logically, both membership and vagueness have values that are infinite and multivalued on $[0, 1]$, but in practice they tend to be finite and multivalued. (In Fig. 11.8, membership is quantified into five values and vagueness into three.) Determining the number of values is problem number (2) from the beginning of this section. Let us consider this an optimization problem using Shannon's concept of entropy.[8]

Let there be n number of values for choices, and let them be arranged in equal divisions on interval $[0, 1]$. In other words, for each value α_i, we get

$$\alpha_i = (i - 1)/(n - 1); \qquad i = 1, 2, \ldots, n, \tag{11.2}$$

and we write the frequency of choice p_i:

$$\sum_{i=1}^{n} p_i = 1; \qquad p_i \geq 0. \tag{11.3}$$

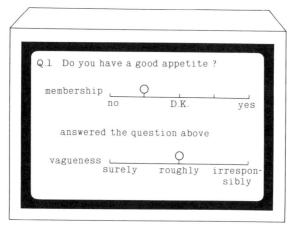

Fig. 11.8.   Examples of Display for MMS New Data Entry System

The entropy H for this condition is given by

$$H = -\sum_{i=1}^{n} \{\alpha_i p_i \log_2 \alpha_i p_i + (1 - \alpha_i) p_i \log_2 (1 - \alpha_i) p_i\}. \tag{11.4}$$

H/n, the dividing of H by the number of conditions n, gives the average amount of information for one choice, and that value changes according to the frequency of appearance pi for each α_i. If we find the maximum value for H/n, $(H/n)_{max}$, based on the restrictions of Equation (11.3), it gives us the average maximum amount of information for each choice α_i, when there are n number of choices. This can be found simply using Lagrange's method of the undetermined coefficient, and we obtain the results shown in Fig. 11.9. In other words, $n = 3$ is optimal, and we can see that the order from there goes $n = 4, 5,$ 6 and that when we get to $n = 7$ it is worse than $n = 2$. Therefore, we verify the propriety of a three-choice multiple choice answer, yes and no with the addition of D. K., and we confirm the fact that if we have seven or more choices, reproducibility is more of a problem than for just two choices, yes and no.

In general, people take "moderately" good actions, but we cannot always hope for the optimal case. Accordingly, we do not immediately choose to use three choices from Fig. 11.9, but we make sure to consider the optimal order of descent, $n = 3, 4, 5, 6, 2, 7, \ldots$. Actually, using psychological tests and human pattern discrimination experiments, experience tells us that it is good to express membership with five choices and vagueness with three, as in Fig. 11.8.

When we know that there is a clear yes/no answer, one bit, two choices, is enough. However, when we cannot help but consider ambiguous cases, we must allow three bits ($= 8 = 5 + 3$ conditions) for membership and vagueness. We can obtain better results than with the data entry system used up to now by utilizing points like (1) the ability to remove the "suspiciousness" from cases in

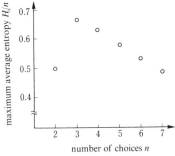

Fig. 11.9. Relation between Number of Choices and Maximum Average Entropy

which a person is not really sure but enters data anyway and (2) the ability to give low-priority rankings to information with large vagueness inputs when inference is carried out.

11.3   MULTISTAGE DECISION MAKING USING FUZZY DYNAMIC PROGRAMMING

The reason for making mathematical programming fuzzy is to allow the model of the object or evaluation to have ambiguity and to come up with a solution that seems good. Since this is close to the way people think, they readily accede to using it. Fuzzy dynamic programming is especially important because it includes all the problematic points of decision making. Parts of the model that can include fuzziness include "state variables," "control variables and external disturbance," "state transitions," "constraints," "decision point" and "truth value of rules." However, the basic problem structure cannot be made fuzzy. If fuzziness is introduced to that extent, it ceases to be mathematical programming. On the other hand, there are many parts of the evaluation side that can also accept fuzziness, aspects such as "evaluation value," "order of preference," "optimal point," "multiobjective tradeoffs" and "final state."

In general, what is called fuzzy dynamic programming has (1) system state, (2) state transitions, (3) constraints, (4) final state, or (5) evaluation values expressed by fuzzy sets. In addition to these, (6) decision points (time and place) are extremely important in real problems. Since this is an independent variable, it may seem strange to introduce fuzziness into it, but decisions points are not always fixed in people's planning activities. Moreover, in standard fuzzy calculations, the ambiguity of the states accelerates as the number of stages increases, and the usefulness of dynamic programming is lost; therefore, we must consider the introduction of fuzziness into states and decision points in terms of groups when modeling. In order to understand the usefulness of introducing fuzziness, it is best to consider first a conversion into ordinary sets of a discrete multistage decision process like that in Fig. 11.10. In the figure, the "points" are system states, the "lines" are transitions, and the "choice of line" is the control (decision making). The final states and constraints can be expressed as "possible destinations." In addition, evaluation is not shown in this figure, but it is given by a function of state and control.

The introduction of fuzziness into (1), (3), (4), and (6) means establishing sets of nodal points, as in Fig. 11.11, and then making them into fuzzy sets. In addition, (2) corresponds to fuzzy mapping and (5) to fuzzy order relations.

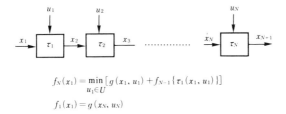

$$f_N(x_1) = \min_{u_1 \in U} \left[g(x_1, u_1) + f_{N-1}\{\tau_1(x_1, u_1)\} \right]$$

$$f_1(x_1) = g(x_N, u_N)$$

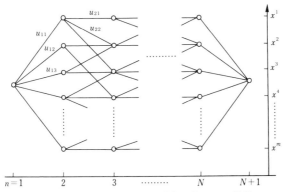

Fig. 11.10. Ordinary Multistage Decision Process (Microstructure)

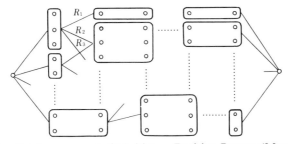

Fig. 11.11. Set Arrangement of Multistage Decision Process (Macrostructure)

Fuzzy transition was originally the expression of the laws of cause and effect, which are based on human experience, in the form of state propositions ("if–then" rules). Fuzzy relations, fuzzy inference, and the combination of ordinary equation and fuzzy numbers can be used as models of fuzzy transition.

In this way there are many possible concepts for introducing fuzziness into the system taken as an object, but since the ambiguities of (1)–(6) have an overall connection, we should not indiscreetly introduce fuzziness. Because the problems with fuzzy dynamic programming are many and varied, it is necessary to try to penetrate the essence of the problem and extract only the important points. The first is that the "if–then" rules that express state transitions do not normally include time explicitly. Therefore, the number of stages is not clear; if it is increased, the ambiguousness of the states increases, and if it is decreased, the truth of the laws of cause and effect becomes ambiguous. The second is that because, for each stage, we must calculate not only the optimal decision, but also quasi-optimal decisions, it can be said that we fall into "the curse of dimensions" and the good aspects of mathematical programming are not brought about. The third is that, in order to give rise to "the principle of optimality" in fuzzy dynamic programming, it must be brought out in definite microsystems, as in Fig. 11.10, before the formation of sets. In contrast to this condition with a system of sets like that in Fig. 11.11, even if the state transitions and evaluation functions are Markovian, one cannot guarantee that the principle of optimality will arise when they are analyzed on a micro level. Since we must think of the transition of the elements of fuzzy sets one at a time, dealing with them is the same as dealing with a microsystem. In other words, we have to start with something like Fig. 11.10 and build our model from there if we want to use dynamic programming calculations. The fourth is that constraints, evaluation, goals, etc., have tradeoff relations with each other, and fuzzy dynamic programming turns out to be one kind of multiobjective optimalization problem.

  Since any one of these is a rather large problem, there are times when, as mentioned before, they offset the usefulness of introducing fuzziness. In order to deal with this, it is best to reduce the number of decision stages, but then accuracy decreases, so here we must consider hierarchical decision making. There are two methods for this. One of them, appropriate for planning problems, is a method in which, first, as is shown in Fig. 11.12, a number of midway goals are determined using macro-level fuzzy dynamic programming; and next they are tied together by micro-level dynamic programming. With this method we cannot obtain a strictly optimal solution, but with ambiguous problems expediency is more important than mathematical strictness. The other method is aimed more at optimal control problems than at planning, but as is shown in Fig. 11.13, it is a method in which a number of optimal controls are determined for the conditions beforehand and are switched back and forth depending on the circumstances. It is not always necessary for the lower-level optimal control to be fuzzy, but the rules for the switching policy (metarules) probably turn out to be fuzzy. On the lower level, we might have control that places weight on economical operation under normal circumstances, but when a large disturbance comes in, it switches to a level of control with maximum safety.

macro-decision process

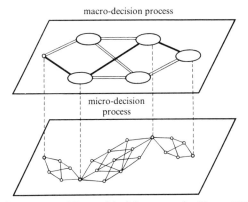

micro-decision
process

Fig. 11.12.   Hierarchical Structure for Fuzzy DP

information   mota rules

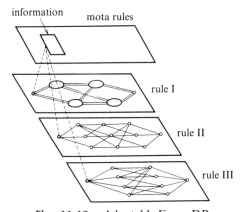

rule I

rule II

rule III

Fig. 11.13.   Adaptable Fuzzy DP

Next, as a concrete example of fuzzy dynamic programming, let us consider a course determination problem for a sail-assisted ship. Figure 11.14 shows a 26,000-ton bulk freighter for which the operation of the sails is under automatic computer control. Let us say the sails are to be used for a maximum fuel savings of 30%. The speed of this type of ship naturally varies with wind speed and direction. Figure 11.15 shows the relationship between wind direction and the speed of the ship for a wind velocity of 30 kt. We shall assume that when wind speeds exceed 40 kt, the sails are not used for reasons of safety and that when the ship speed drops below 6.6 kt., an auxiliary engine begins to work.

There is much observational data about ocean wind directions and speeds, and probability distributions over the entire year are known for points on each ocean. We now divide the ocean into a grid, and using this data and

Fig. 11.14. SWIFT WINGS (The Japan Shipbuilding Industry Foundation, Japan Marine Machinery Development Association)

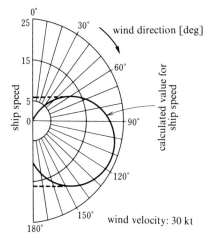

Fig. 11.15. Example of Relation between Wind Velocity and Ship Speed

probabilistic dynamic programming, we can determine courses that minimalize the expected values for fuel consumption and time required. However, the calculation time using probabilistic dynamic programming is very long, and over and above this problem, current values are probably more important for actually determining the ship's path than annual data. In addition, forecasts of storms and the captain's experience in means of avoiding them come into play. To put it another way, the experience and perceptions of a skilled person are more important than statistical data. Therefore, fuzzy dynamic programming is more appropriate than probabilistic dynamic programming for an optimal course problem.

We let, current wind speeds and directions for each division of the grid, obtained from weather maps and other sources, be expressed by fuzzy numbers. We let the current value be the vertex of the membership function for fuzzy wind direction and let it take the shape of an equilateral triangle with a ± 30 degree spread. The fuzzy wind speed takes a similar form with a ± 5 kt. spread. By means of these we can calculate fuel consumption and time required when the ship moves from any one point on the grid to any of three others. Since this example is relatively simple, it is not necessary to use hierarchical dynamic programming. Using a microcomputer, the calculation time with probabilistic dynamic programming is 30 minutes; and with fuzzy dynamic programming it is about 30 seconds.

We can think of both minimal fuel cost and minimal time required as evaluation functions for the optimal course, but each of these is a fuzzy number, so there are six types of policies, such as taking the value at the vertex, or taking the maximum value after an α-cut, or the minimum value for the same. The choice of α is of course one of the possible policies, depending on the aims of the shipping company or the captain. Fig. 11.16 is an example of the results of using fuzzy dynamic programming, and it shows optimal

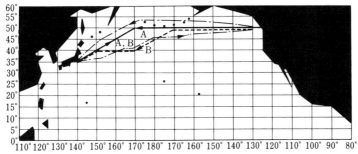

Fig. 11.16. Optimal Routes between Tokyo and Vancouver

Table 11.2. Elapsed Time and Fuel Consumption for Optimal Course:
Vancouver → Tokyo

Course	Distance [km]	Elapsed Time (h)			Fuel Consumption (*l*)		
		Min. Value	Vertex (Value)	Max. Value	Min. Value	Vertex (Value)	Max. Value
A	7 844.51	154.79	544.27	641.77	0	11 557.1	36 974.1
B	8 118.62	160.29	554.29	664.20	0	11 114.0	38 266.2

Tokyo → Vancouver

Course	Distance [km]	Elapsed Time (h)			Fuel Consumption (*l*)		
		Min. Value	Vertex (Value)	Max. Value	Min. Value	Vertex (Value)	Max. Value
A, B	7 844.51	154.74	469.30	641.77	0	7 505.05	36 974.5

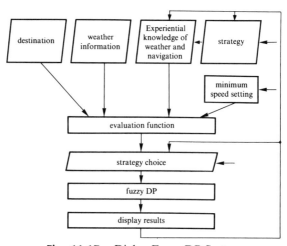

Fig. 11.17. Dialog Fuzzy DP System

courses from Vancouver to Tokyo with line A and dashed line B.[11] A is the
case using minimalization of the max value of fuzzy evaluation, B is the case
using minimization of the vertex values.* In the same way we determine the
optimal courses from Tokyo to Vancouver, and in this case A and B turn out

---

* In this example, the courses for minimal fuel consumption and minimal time inadvertently
turned out to be the same. In addition the courses for the minimalization of the max and min
values also turned out the same.

to be the same as A in Fig. 11.16. The evaluation functions for these cases are shown in Table 11.2. In addition, the actual course taken by the ship in Fig. 11.14 is shown by dash/dot lines in Fig. 11.16.

Up to this point we have been dealing with course plans prior to departure, but in reality, because of weather changes, load conditions, economy, and safety, it is often necessary to change course. If we consider these factors, fuzzy dynamic programming must be a dialogue system with easy input and output. In addition, the output must not only show the course, but also give estimated time of arrival and estimated fuel consumption in fuzzy numbers. Therefore, the complete structure must be a human-machine system like that in Fig. 11.17. The horizontal arrow in the figure shows the place for human intervention.

REFERENCES

(1) Swain, A. D., and Guttman, H. E., *Handbook of Human Reliability Analysis, with Emphasis on Nuclear Power Plant Applications*, NUEREG/CR-1278, NRC (1980).

(2) Terano, T., Murayama, Y., and Akiyama, N., Human Reliability and Evaluation of Man-Machine Systems, *Automatica*, 19 (1983).

(3) Terano, T., Murayama, Y., and Akiyama, N., "Human Reliability in Parallel Jobs," in Bezdek, J., ed., *The Analysis of Fuzzy Information* Vol. II, CRC Press (1986).

(4) Terano, T., Masui, S., Murayama, Y., and Akiyama, N., "A Structural Model of Human Reliability Using Fuzzy Sets, "Trans. SICE, **23**, 1 (1987) (in Japanese).

(5) Kohno, S., "A Study of Human Reliability," Bachelor thesis, Hosei University (1985) (in Japanese).

(6) Czogala, E., *Probabilistic Sets in Decision Making and Control*, Verlag TUV Reinland (1984).

(7) Hirota, K., *Extended Fuzzy Expression of Probabilistic Sets*, ed. Gupta, M. M., pp. 201–214, North-Holland (1979).

(8) Hirota, K., *Ambiguity Based on the Concept of Subjective Entropy*, Gupta, M. M., and Sanchez, E., eds., pp. 29–40, North-Holland (1982).

(9) Terano, T., "Some Problems of Fuzzy Dynamic Programming," paper presented at 1st IFSA Conference, Palma (July 1985).

(10) Terano, T., Murayama, Y., Kikuchi, M., Course Optimization for Sail Assisted Ship Using Fuzzy DP, Proceedings, Annual Conference of SICE (1985) (in Japanese).

(11) Nomoto, N., "Optimum Route for Sail Assisted Ship—An Application of Fuzzy DP," Bachelor thesis, Hosei University (1986) (in Japanese).

(12) Esogbue, A. O., and Bellman, R. E., "Contribution to Fuzzy Dynamic Programming," 2nd World Conference on Mathematics at the Service of Man, Las Palmas Spain (1982).

(13) Baldwin, J., and Pilsworth, B. W., "Dynamic Programming for Fuzzy Systems with Fuzzy Environment," *Journal of Mathematical Analysis and Applications*, **85**, 1–23 (1982).

(14) Vira, J., "Fuzzy Expectation Values in Multi-Stage Optimization," *Fuzzy Sets and Systems*, **6** (1981).

Chapter 12

ROBOTS

There are fewer examples of tests using control methods based on fuzzy inference for robot control than there are of tests using them for plant or transport control. However, at the beginning of the 1980s, progress was made in fitting industrial robots that simply repeated actions they were instructed to do with various kinds of sensors for perceptive functions and human-machine information exchange. By adding artificial intelligence techniques, the vigorous development of intelligent robots using fuzzy control started. Of all of the possible examples we will introduce three here: a path-determining robot that makes use of a moving robot with sensors, an industrial robot that can grasp objects moving along a conveyor belt using a CCD camera, and an arm robot, made from an industrial arm robot and touch sensors, that infers the position of an object.

12.1 PATH-JUDGING ROBOT

The origin of the word *robot* is in Czech writer Karel Capek's play "Rossum's Universal Robots" (1920), and with the advent of computers there have been expectations of mechanical dolls that are just like humans. Although the technology of the 1960s failed to create robots, in the 1970s rapid progress was made on industrial robots that centered on repetition of given actions, and with the coming of the 1980s the investigation of making robots intelligent spawned the expression "intelligent robot." In general, machines with various sensors and other features that allow them to have sensory functions for perceiving the circumstances and the environment around them, with intelligent information-processing functions related to artificial intelligence, with motor functions that provide for appropriate movements, and with human-machine information exchange functions that provide for mutual understanding of intent with people are called intelligent robots. The hardware for the intelligent parts is a computer centered on a microprocessor.

If we take a look at the history of computers, we see that in 1975 the American company MITS announced the ALTAIR computer kit, which was based on the Intel 80, and the first personal computer, a computer meant for private ownership, made its appearance. This attracted the attention of many people (1975–76), and the APPLE II, PET, TRS-80, etc., opened the market (1977–1978). There was great growth in the industrial base (1978–82), and now what was once the monopoly of a few people in research institutes and universities has become the high-level compact computer used freely by any individual.

In the case of robots, privately owned robots are known as personal robots. The American company Health Co. announced the production in 1981 of HERO I, a kit for a mobile robot equipped with sensors and designed for private ownership. Soon after, the invention attracted the attention of many people (1981–82) and the RB5X from R. B. Robot, Inc. and the MTR 801 from Mitsubishi (1982–84) were announced.

Here we will describe one of the experiments aimed at providing the first personal robot, HERO I (Fig. 12.1) with intelligence. It is a path-judgment experiment. Micromouse competition is a well known path-judgment area, but it differs from the micromouse setup, as follows.

Setup

Relying on a rough hand-drawn map, the robot somehow arrives at a destination in a building it has never visited. In other words, the object is to arrive at the destination using incomplete information, without placing too much emphasis on a little wasted action or time.

There have been few studies of this kind of problem using sensor-equipped robots, and first we will give a simple description of the specifications of HERO I, following the four conditions for intelligent robots.

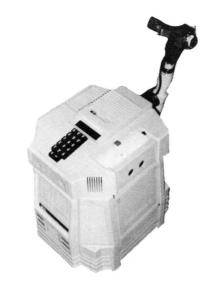

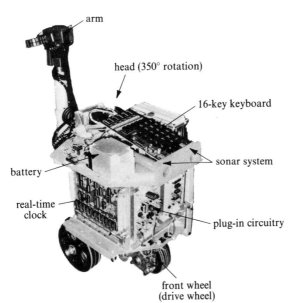

Fig. 12.1. Mobile Robot with Sensors: HERO 1

Sensory Functions

It has sound sensor (8-bit (256-step) sound discrimination in the audible range for humans); light sensor (8-bit (256-step) brightness in the visible range for humans); motion detector (ability to perceive the motion of moving objects within 4.5 meters); ultrasonic distance measurement (measurement of the distance of objects from itself within 2.4 meters).

Intelligent information Processing Functions

It has an 8-bit MC 6808 microprocessor from the American company Motorola with 8K-byte ROM and 4K-byte RAM. The program is in the form of an interpreter called *machine language* or *robot language*, which is written using a number of commands.

Motor Functions

It is capable of self-propulsion, turning its head and extending and retracting its arm (and wrist). It moves by means of three wheels, one in front and two in back, and the single front wheel provides steering and is the drive wheel.

Human-Machine Information Exchange Functions

It has a conversation function using a phonemic composition method, and a code-based teaching pendant that makes use of a six-digit seven-segment LED with decimal point.

This kind of intelligent robot is endowed with the minimum level of necessary functions, and if we compare them with an average person, they are extremely poor. Accordingly, when we say path judgment, it is as if a person who is hard of hearing and has poor eyesight tries to travel in a wheelchair using a staff to feel his way. A step would cause trouble; there is no image processing function, and voice discrimination is impossible. Therefore, the restrictions for the course from the beginning to the end are as follows.

Restrictions

There must be walls on both sides, a level floor and a limited number of intersections. In addition, we limit the intersections to orthogonal ones and $8 (=2^3)$ shapes, such as a cross or a T.

The form of laboratory that satisfied these requirements was somewhat like an eel's den, a university building with long corridors and intersections of various shapes here and there. However, since we had to assume a lack of obstructions, such as people walking in the halls, the experiment was conducted on holidays and at night.

Following is the information given to HERO I about the path from start to finish, which corresponded to the hand-drawn map. Even when a person who knows the correct path conscientiously draws a map and gives instructions, there are many cases in which information, especially the distances from one intersection to the next turn (we will call this straight-line distance), even if given by expressions such as "about 100 meters," are actually quite inaccurate. Therefore, we let the path information contain accurate information of the forms of the intersections and the directions to be traveled, but let the straight-line distance information be ambiguous. The joint use of the vagueness concept described in 11.2 was employed as the method of expressing this ambiguous straight-line distance information. Depending on whether the "about" of, for example, "about 30 meters" is confident or lacks confidence, the permissible range (the range in which the actual distance exists) for "30 meters" is either made smaller or larger. Therefore, we let the straight-line distance be expressed by a numerical value (in meters) + vagueness (three values: 0, 0.5, 1). In this way the information about the course from beginning to end was given by means of a limited number of groups of ambiguous straight-line distance information, accurate information about the intersection ahead, and accurate information about the direction of travel.

Following this kind of information about the course, HERO I was sent on its own power toward the goal. The ultrasonic search range for the next intersection corresponded to the vagueness value and was set up as shown in Fig. 12.2. The program was written in machine language and robot language,[1] and the results of the experiment were that the robot was able to travel the total length of the course, which was about 1 km (this restriction mainly dictated by the charge of the battery and the structure of the building) on its own power, and to arrive at the goal. Actually, the course was completed solving many small problems one by one, beginning with revisions due to the properties of the sensors, but these are small problems from the point of view of fuzzy theory, so we will refrain from discussing them any further.

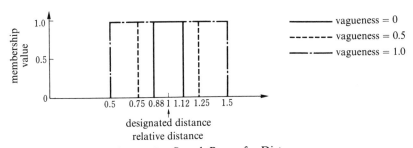

Fig. 12.2.   Search Range for Distance

12.2   OBJECT-GRASPING ROBOT

Following are the results of the construction of a real-time system for grasping
an object carried by a conveyer belt and moving at variable speeds, using a
multijoint vertical articulated robot with five degrees of freedom and a CCD
camera. The expression of ambiguous information was done using membership functions, which have been used all along in fuzzy theory, along with the
concept of vagueness from extended fuzzy expressions as described in 11.2. A
simplified system was planned that would allow real-time control of the
complete system, which included two industrial robots and a CCD camera
image processing system, using a single all-purpose 16-bit microprocessor.

A general view of the system is shown in Fig. 12.3. It is made up of two
multijoint vertical articulated robots (Mitsubishi RM501) with five degrees of
freedom, one CCD camera (Sony XC-37), an image frame memory with a
resolution of 256×256 pixels with 64 gradations (6 bits) gray scale (Edec ED-
1161) (with two frames worth of RAM, total 128K built in), one monitor for
showing the camera image, and one personal computer (NEC PC-9801) (8086
& 8087 CPU, 5MHz clock, 384 K-byte RAM) with its peripheral equipment
(keyboard, monitor, 8-inch floppy disk drives) connected to the robots and
image memory to control the whole system. In addition, the objects were
whitish things with dimensions of $5 \times 5 \times 5$ cm or less that would fit the grip

Fig. 12.3.   Robot System for Grasping Objects

of the robot hand, and they were moved by means of an industrial variable speed mini-conveyer belt (Sanwa Conveyor). One robot arm was used as a base for the camera to move in correspondence with the movement of the object.

The program was written using N 88-Japanese BASIC (86) (MS-DOS version) and assembler language (the basic part being composed of 700 steps, and the assembler part of 2K bytes including data). The overall flow of the program was BASIC, and the fuzzy inference section was accelerated using assembler language. The algorithm is shown in flow-chart form in Fig. 12.4. The whole is divided into six blocks: observation, quantification, inference, interpretation, (robot) control, and grasping. In what follows, we will give a simple explanation of each.

Observation Block

The moving object is observed and necessary information extracted. The necessary information is the speed of the object and the distance between it and the hand, and since real-time high-speed processing is carried out based on

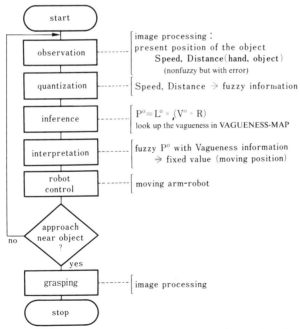

Fig. 12.4.   Action Algorithm for Object Grasping Robot System

a frame difference method, the information passed on to the following blocks contains a large amount of error.

Quantification Block

Fuzzy information is constructed so as to take into account error as well as the measurements extracted by the observation block. Let us consider a model in which the speed of the object is changing slowly and continuously.

First, for the speed of the object V_O in fuzzy terms, the change in distance moved between frames Δd is divided by the frame time Δt and apportioned among the elements of a standardized support set $\{0, 1, \ldots, 14\}$. If Δt is small, the reliability of the calculated speed falls, so we let the membership function of the element assigned be 0.8 and the values of other elements each be 90% of the value for that element for the previous time. When Δt is large, we let the value of the membership function of the mapped element be 1.0, and the other elements each be 60% of the previous value. In this way the speed information of the previous step is reflected in the current information about speed.

Next, for the distance L_O between the hand and the object in fuzzy terms, the measured values are first apportioned to the various elements of a standardized support $\{0, 1, \ldots, 16\}$. The membership function is determined considering the relative positions of the hand and the object. When the hand is on the side toward which the object is moving, we let the membership value of the value assigned be 0.6, that of the next smaller element be 1.0, and that of the next smaller one be 0.4. When the hand is in the vicinity of the object, we let the membership value for the assigned value be 1.0 and the one smaller and the one larger each be 0.5. When the hand is behind the object (on the opposite side from the direction of travel) we let the membership value for the assigned value be 0.2, that for the next larger be 1.0, and that for the next larger be 0.5.

Inference Block

Based on the two pieces of ambiguous information (V_O, L_O) found in the quantification block, inference concerning the (anticipated) position P_O is carried out, and the vagueness value is determined.

The standardized fuzzy labels in Table 12.1 were set up for the speed of the object and the hand–object distance V, which are given in the inference propositions, and for position P, which is obtained as the result of predicting the position of the object from its present position to the next step. The re-

Table 12.1.   Fuzzy Labels for V, L, and P

	low												high		
	0	1	2	3	4	5	6	7	8	9	10	11	12	13	14
V_1	1	1	1	0.5	0.1	0	0	0	0	0	0	0	0	0	0
V_2	0	0.2	0.4	0.8	1	0.8	0.4	0.2	0	0	0	0	0	0	0
V_3	0	0	0	0	0	0.2	0.4	0.8	1	0.8	0.4	0.2	0	0	0
V_4	0	0	0	0	0	0	0	0	0	0.2	0.4	0.8	1	1	1

(a) Fuzzy labels for velocity (V)

	near														far		
	0	1	2	3	4	5	6	7	8	9	10	11	12	13	14	15	16
L_1	1	0.8	0.2	0	0	0	0	0	0	0	0	0	0	0	0	0	0
L_2	0.1	0.6	1	0.6	0.1	0	0	0	0	0	0	0	0	0	0	0	0
L_3	0	0	0.1	0.6	1	0.6	0.1	0	0	0	0	0	0	0	0	0	0
L_4	0	0	0	0	0.1	0.6	1	1	1	0.6	0.1	0	0	0	0	0	0
L_5	0	0	0	0	0	0	0	0	0.1	0.6	1	1	1	0.6	0.1	0	0
L_6	0	0	0	0	0	0	0	0	0	0	0	0	0.1	0.6	1	1	1

(b) Fuzzy labels for robot–object distance (L)

	small movement														large movement		
	0	1	2	3	4	5	6	7	8	9	10	11	12	13	14	15	16
P_1	1	0	0	0	0	0	0	0	0	0	0	0	0	0	0	0	0
P_2	0.1	0.6	1	0.6	0.1	0	0	0	0	0	0	0	0	0	0	0	0
P_3	0	0.2	1	0.2	0	0	0	0	0	0	0	0	0	0	0	0	0
P_4	0	0	1	0.6	1	0.6	0.1	0	0	0	0	0	0	0	0	0	0
P_5	0	0	0	0	0.1	0.6	1	0.6	0.1	0	0	0	0	0	0	0	0
P_6	0	0	0	0	0	0	0.1	0.6	1	0.6	0.1	0	0	0	0	0	0
P_7	0	0	0	0	0	0	0	0	0.1	0.6	1	1	0.6	0.1	0	0	0
P_8	0	0	0	0	0	0	0	0	0	0	0	0.1	0.6	1	1	1	1

(c) Fuzzy labels for (inferred) distance to be moved (P)

sults of inference P_k for this kind of condition (V_i, L_j) are obtained by means of the control rules shown in the 24-part compound proposition of Equation 12.1 and are expressed in table form in Table 12.2.

$$\text{IF } V \text{ is } V_1 \text{ AND } L \text{ is } L_1 \text{ THEN } P \text{ is } P_1 \text{ ELSE}$$

$$\text{IF } V \text{ is } V_1 \text{ AND } L \text{ is } L_2 \text{ THEN } P \text{ is } P_1 \text{ ELSE}$$

$$\vdots \qquad\qquad (12.1)$$

$$\text{IF } V \text{ is } V_4 \text{ AND } L \text{ is } L_6 \text{ THEN } P \text{ is } P_8$$

Table 12.2.   Fuzzy Inference Rules

		near ---------- far					
		L_1	L_2	L_3	L_4	L_5	L_6
low	V_1	P_1	P_1	P_1	P_1	P_1	P_1
↑	V_2	P_1	P_2	P_2	P_3	P_4	P_5
↓	V_3	P_1	P_2	P_4	P_5	P_6	P_7
high	V_4	P_1	P_3	P_5	P_6	P_7	P_8

small movement ←----------------- large movement

$$P_1 \quad P_2 \quad P_3 \quad P_4 \quad P_5 \quad P_6 \quad P_7 \quad P_8$$

The fuzzy relation R obtained from this compound proposition is a $15 \times 17 \times 17$ three-dimensional configuration (given by products of membership functions), and for real-time high-speed processing is calculated beforehand and placed in memory.

In this way control output P_O is obtained as a fuzzy variable. Furthermore, the concept of vagueness is introduced in order to allow consideration of the accuracy of inferred value P_O. This value is found for the two conditions (V_O, L_O) directly from a vagueness table (Table 12.3), which is given beforehand.

Interpretation Block

The output of the inference block is fuzzy information given by membership functions into which vagueness has been introduced. In order actually to make the robot move, this must be interpreted as a fixed number.

The method used for the interpretation of fuzzy output P_O as fixed number p was a random choice method using weighted random numbers of membership values from a range above the membership threshold value. (In addition, the method using the center of gravity was programmed, and it was possible to use

Table 12.3.   Vagueness Map

	L_1	L_2	L_3	L_4	L_5	L_6
V_1	0	0	0	0	0	0
V_2	0	0	0	0.2	0.3	0.4
V_3	0	0.2	0.2	0.3	0.4	0.5
V_4	0	0.3	0.3	0.4	0.5	0.5

it.) Next, conversion to actual coordinates was done. In consideration of safety in the moving of the hand, a device in which a vagueness value of 1 was given as a superscript to the calculated value for the actual distance to be moved was used.

(Robot) Control Block

The hand is moved to the coordinates found in the interpretation block. The calculations for the angle patterns to which each joint motor is turned are done as standard geometrical calculations.

Grasping Block

When the hand is close enough to the object, the hand is actually closed and the object grasped.

Using the above program, it was possible to construct a robot system that could grasp objects with varying speeds as they moved along the conveyor belt. The locus of the hand did not always follow the optimal path, and based on the fuzzy information, there were times when there was some wasted motion. However, it was possible to grasp the object fairly accurately in about three to nine steps. The important points are that it was acceptable to have ambiguity in the necessary information for the robot, yet using only a 16-bit all-purpose computer, an object moving at a variable speed could be grasped in real time while image processing was being carried out. A speed for the moving object of up to about five cm/s maximum could be imitated by the setup in Fig. 12.3.

12.3 PLACEMENT INFERENCE ROBOT[2]

When dealing with qualitative (ambiguous) commands given by others, people use common sense to solve problems. Therefore, we will explain the workings of an arm robot (five degrees of freedom) that receives qualitative commands from people and, using touch sensors, performs the same types of activity that people do to infer the placement of an object. This placement-inference robot conducts a search for an object placed on a stand or tabletop based on information presented by people and conducts activities as people would, so its movement differs from the random searches used up to now. The robot itself must be able to understand the qualitative commands in the same way people do and must have circumstantial discrimination and internal knowledge.

12.3.1   Hierarchy of Decision Rules

Qualitative commands are vague or incomplete. Nonetheless, given this type of command, people can process problems skillfully. Thus we would like to have robots that can understand vague and incomplete commands and take the actions that people would take. However, it is difficult to make rules for the actions people carry out for vague or incomplete commands such as "look for the object" or "look on the right side" (when the object is actually on the left side). Therefore, we list possible human actions and make a hierarchical structure in order to understand commands in terms of a simple production system with a tree structure. The robot for this experiment has touch sensors only in one hand, so we consider the groping search activities of humans in a dark room when given incomplete information. These human activities are made into rules, expressed in "if – then" form, and arranged in the hierarchical structure shown in Fig. 12.5. The rules are divided into four stages: evaluation of the reliability of command, determination of action, determination of action range, and recognition of the object. The rules of the hierarchical structure are "if – then" rules like

C_3: if it is not found after a broadening of the search area, then the truth value of the given indication is more or less false;

R_3: if the indication is false, then search a place other than that indicated;

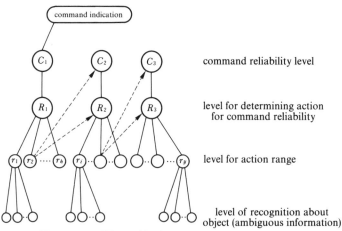

Fig. 12.5.   Hierarchical Structure of Decision Rules

r_{20}: if the first indication is "right," and the result is "search an area other than that indicated" then the search range is "wide left" given by a fuzzy number;

when, for example, there is a mistaken command. When the object cannot be found using the lowest rules, it is necessary to backtrack. This backtracking is carried out using metarules, constructed using Fault Tree Analysis (FTA). If we use the FTA in Fig. 12.6, the cause-and-effect relations for when the object cannot be found become clear, and metarules can easily be constructed. Fig. 12.7 shows a placement-inference flowchart. The command is given; passing through the rules for judgment of reliability, the action rules are determined, and with understanding of the circumstances of the command, the membership function for the fuzzy action range is determined.

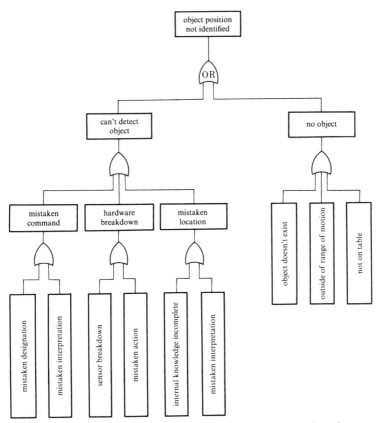

Fig. 12.6.   Fault Tree Analysis for Object Placement Search

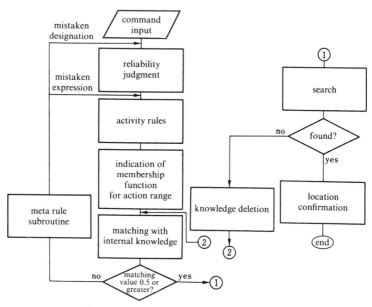

Fig. 12.7.   Location Inference Flowchart

12.3.2   Circumstantial Understanding of the Command

After the estimation of the intent of the command is complete, the circumstantial understanding of the robot is expressed by a fuzzy range, the circumstances being qualitative for vague or incomplete commands. The membership function of the fuzzy range for this circumstantial understanding is determined using default inference that takes into consideration human actions. For example, in a case in which no location is indicated and the command is simply "look for the object," we let a silent understanding exist that the object is on the stand and express, with fuzzy numbers obtained from the default inference, the common-sense activities people would carry out. That is, when people search for an object, they probably start from the vicinity of the center of the stand. Therefore, we let a silent understanding that the search should begin from the near the center of the stand exist, and we let the circumstantial understanding be that the "object" is in "the vicinity of the center from right to left" and in "the vicinity of the center from back to front." We then introduce the membership functions for circumstantial understanding C_x (vicinity of the center from left to right) and C_y (vicinity of the center from back to front). On the contrary, when the location of the object is indicated, the membership function for circumstantial understanding is constructed in agreement with what is indicated. For instance, the membership function of the indication "search for the object on the right" is shown by the dotted lines at the right and bottom of Fig. 12.8. In spite of the fact that the circumstances for commands are understood in terms of fuzzy numbers, comparison of all the positions on the stand with them takes time. Thus, the internal knowledge of the positions searched is also introduced in terms of fuzzy numbers.

12.3.3 Internal Knowledge or "Search Points"

This placement inference robot does not require strict accuracy, and it is enough if the goal is attained for a given qualitative command. Therefore, the robot's internal knowledge expresses the minimal search points for the range in which it is possible to attain the goal, each in terms of a fuzzy number, and using a small amount of internal knowledge it is able to process a large amount of information. This internal knowledge is determined by the size and shape of the touch sensors. Here the shape of the sensors is a hollow cone, and with internal knowledge of only 22 search points, the entire surface of the stand can be searched. In other words, even without the large number of points used in random searches up to now, the object can be searched out using inference, if there are just 22 search points. Since the touch sensors are hollow cones, the membership functions turn out to be the equilateral triangles drawn at the bottom and side of the stand in Fig. 12.8. However, if the sensor were a round

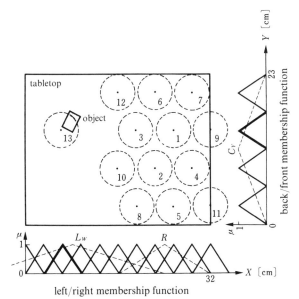

Fig. 12.8. Search by Matching Circumstantial Understanding and Internal Knowledge (When There is a Mistaken Designation)

plate, the sensor would work when any part just barely touched the object, and it would be impossible to infer the position of the object. This means that the shape of the touch sensor should be based on fuzzy numbers.

12.3.4   Inference of the Object's Position

Since Fig. 12.8 shows the conditions of a search for the command "search for an object on the right," matching between the fuzzy set for circumstantial understanding and that for internal knowledge is carried out. There are many studies of methods for this matching, but here the method of converse truth qualification, which uses truth values, is used.[3] In other words, when fuzzy sets A and B, which express ambiguous circumstantial understanding and ambiguous knowledge, are given, the degree of matching of A and B is numerically defined by the following.

$$t = N(A/B) = \tfrac{1}{2}\{\sup(A \cap B) + \inf(A \cup \daleth B)\}. \qquad (12.2)$$

The search for the object is conducted in the order of the size of matching value t, which is the minimum matching value of $(R$ and $P)$ and $(C_y$ and $P)$, where R, C_y, and P are the fuzzy sets of circumstantial understanding, silent understanding (front and back), and internal knowledge, respectively, as shown in Fig. 12.8. Here, the first location searched is 1, and others are searched in order, 2, 3, ... In this example, the right side was searched, but the object was not there, so the "indication" was judged to be mistaken, and the membership function for circumstantial understanding was changed to "wide left" (L_w), and the result of the search of the left side was the finding of the object. For this search, the search actions that would take place if a person were groping for an object in the dark were hypothesized and then used to establish the rules for the internal knowledge. When the human indication is appropriate, and naturally for incomplete indication also, the robot can make judgments for itself and find the object.

REFERENCES

(1) Hirota, K., *Intelligent Robots with Fuzzy Control* Section 2.4, McGraw-Hill Books (1985) (in Japanese).
(2) Terano, T., and Masui, S., "A Study of Fuzzy Robots," Proceedings of 1st Fuzzy System Symposium, IFSA Japan chapter, Kyoto (1985) (in Japanese).
(3) Gupta, M. M., Tsukamoto, Y., and Nikiforuk, P. N., "Truth Qualification and Numerical Truth Values," *JACCA* (1980).

Chapter 13

IMAGE RECOGNITION

Here we will introduce two examples of applications using fuzzy theory for image pattern recognition. The first concerns visual recognition by a robot's eye. We will describe the methods used for recognizing seven different objects, such as a paper cup and a ball, and determining the direction and distance to them, using a CCD camera and a 16-bit personal computer.

In the second example, we will describe the results of using fuzzy clustering techniques, based on FUZZY ISODATA, in area partitioning by means of texture analysis of LANDSAT images. In addition, we will describe methods of interpretation using entropy for the subtle differences in the results of territorial partitioning (especially boundary sections) that result from different techniques.

13.1 SHAPE RECOGNITION AND DISTANCE/DIRECTION INFORMATION: EXTRACTION USING A CCD CAMERA

Here we will describe image recognition for a robot's eye using an arm robot setup with a CCD camera.

In general, the amount of data to be processed for image recognition is large-scale, and often the processing time is on the order of 10 minutes to an hour. Since the computers that can be used with robots are small-scale ones centered on microprocessors, the amount of memory is also small in most cases. However, if we want to use image processing for a robot's eye, we must process a continuous image data series in real time and can only hope for a process with specific objectives and a limited setting.

Therefore, we deal with images from a standard CCD camera with 256 × 256 image elements, 8 bits per element (frame memory of 64K bytes required for each image) used for the robot's eye. This is used for the eye of a vertical articulated robot with 6-axis simultaneous control, and the computer is a standard 16-bit PC. (The setup of the system is the same as that in Fig. 12.3 in Section 12.2.)

There are two broad types of information necessary when we try to process objects using the robot arm. The first is information about the objects themselves, and what is especially important is the type of object. If it is clear that the object is already known, it is possible to include knowledge of other properties in a database of accessible size and form. Another piece of necessary information is the relative positions of the arm and the object. Since this can be expressed by one three-dimensional vector, the distance information and directional information can be considered separately. If these pieces of information can be suitably extracted, the robot can move its own joints and process the object.

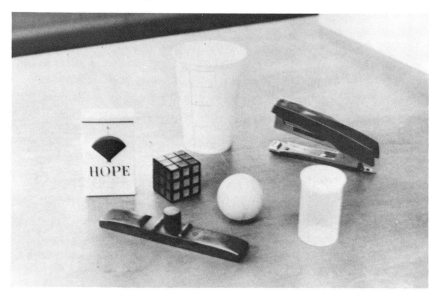

Fig. 13.1.   Seven Types of Objects

There have been numerous studies published on accurate determination of position and shape recognition, but all-purpose methods for real-time processing using small-scale systems are not common. Therefore, a methodology capable of actions using rough information about shape and position would be desirable. Given this kind of setting, fuzzy control shows great advantages. Since several concrete examples of control were discussed in Chapter 12, we will describe shape recognition, ambiguous distance information, and ambiguous directional information, in that order.

Considering properties that cause problems when processing is done by robot hand (heavy, not to be crushed, not to be tipped over, difficult to grasp), the seven commonplace objects in Fig. 13.1 were chosen.

We will now describe the algorithm for distinguishing which of these is the object in the CCD camera input. Basically the method takes the form of extracting several characteristic quantities from the object portion of the input image, comparing these values with information in the knowledge database in the memory, and distinguishing the type of object. Therefore, if the number of objects increases (to about 10), it will take that much more time for processing and it will be necessary to detect a few other characteristic quantities.[1] In addition, there are studies related to artificially intelligent recognition of the shape of any object without previously established limits,[1] but real-time processing is impossible with a small-scale system. After investigating various characteristic quantities, the six characteristic quantities in Fig. 13.2 were used to distinguish the seven types of objects in Fig. 13.1, and we chose to use the

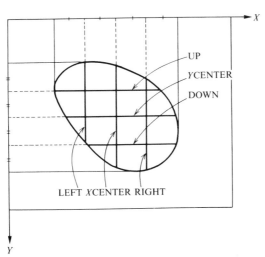

Fig. 13.2. Six Characteristic Quantities for the Objects

algorithm in Fig. 13.3. The six characteristic quantities used are the lengths
where the object exists along the partition lines when the rectangle that
circumscribes the object part of the image, which has undergone a two-value
conversion, has been divided into four vertical and four horizontal partitions;
therefore, they can be found fast enough. The tilde over the equality and
inequality signs in this classification algorithm, which uses the six characteristic quantities, indicates the inclusion of ambiguity with a permissible range
of 10%. Next we will describe the high-speed calculation method for
determining the distance to the object from the CCD camera, which is
attached to the end of the robot's hand. The natural principle that objects that
are closer have larger projections and those farther away have smaller ones is

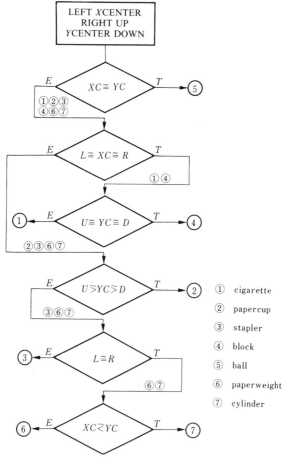

Fig. 13.3.   Main Points of Object Discrimination Algorithm

followed. In other words, if we let the distance from the camera to the object and the area of the object for a two-time experiment be L_0, L_1 and S_0, S_1, respectively, we get

$$L_0^{-2} : L_1^{-2} = S_0 : S_1. \tag{13.1}$$

Since the objects are given as shown in Fig. 13.1, we can find the distance L_1 from the camera to the object using

$$L_1 = \frac{L_0 \sqrt{S_0}}{\sqrt{S_1}} \tag{13.2}$$

by calculating the area of the object S_1 revealed by the camera (image elements), if we put L_0 and S_0 for the standard position of each object into memory. However, there is the possibility of getting a different value for the observed S_1 for the same object even at the same distance, because of the angle from which it is viewed. In other words, the ball is the same when viewed from any angle, but S_1 things, like the cigarettes, will vary greatly if they are viewed from top or front even at the same distance. Therefore, depending on the object, there will be cases when the value for L_1 calculated from Equation (13.2) will be very inaccurate. This can be avoided by standardizing the direction by means of affine conversions, but to do so is a large encumbrance for real-time processing. Therefore, the inaccuracy of the calculated values of L_1 from Equation (13.2) is expressed by the vagueness information shown in Table 13.1, and a method that takes this into consideration is used when the hand is actually moved. In other words, the distance that the hand moves has a membership function like that shown in Fig. 13.4 and can be interpreted as a fuzzy set. (The actual distance the hand travels is a random number weighted by this membership function or by a central axis value.)

Finally, we will describe the method for calculating the rough direction from which the camera views the object. For example, if the object being picked up

Table 13.1. Vagueness for Calculated Distance

Object	Vagueness Value
ball	0
block	0.5
cigarette	1.0
cylinder	0
papercup	0
paperweight	0.5
stapler	0.5

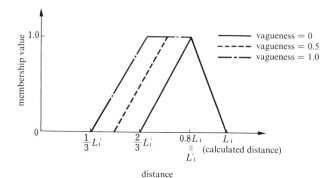

distance

Fig. 13.4. Calculated Distance and Distance to Be Moved (Fuzzy Sets)

by the camera is in the center of the image, we know that the direction of the object is "front," and when it is in the upper center, we know that it is "upper front." Therefore, we divide the image into nine (3 × 3) areas, and depending on which of these areas the object is in, it is judged to be in one of the directions shown in Fig. 13.5. However, it is rare that the object is completely in one of the nine areas, and it is usually spread over several. Therefore, the rough direction is given as that of the area in which the largest portion of the object exists, but the inaccuracy is expressed in terms of vagueness. In other words, a three-level vagueness value is established: if the ratio of the largest part of the area to the total area of the object is less than 0.4, the vagueness is 1.0; if 0.4 or greater up to 0.7, the vagueness is 0.5; and if greater than 0.7, the vagueness is zero. The actual direction in which the hand moves is determined from this ambiguous information inclusive of the vagueness. In order that the object not be lost from the camera's field of vision after the hand moves, a device for the

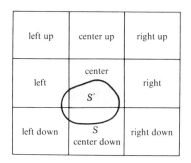

$0 \leqq S'/S < 0.4$: vagueness = 1.0
$0.4 \leqq S'/S \leqq 0.7$: vagueness = 0.5
$0.7 < S'/S \leqq 1.0$: vagueness = 0

Fig. 13.5. Ambiguous Directional Information for the Object

modification of the membership function is employed, but we will leave out the details.[2]

In the above manner the hand faces the object and approaches it using ambiguous information. Since the information from observations is not precise, it is impossible for the hand to be brought to the object in one try; a two- or three-step repeated action is necessary, but this is not much of a problem under real conditions; and conversely, since ambiguous information can generate enough motion, the image processing load is reduced and it is possible to perform real-time activities using a system on the scale of a personal computer. In addition, it is interesting that the precision of the calculated information from each step improves as the object is approached.

13.2 TEXTURE ANALYSIS OF AERIAL PHOTOGRAPHS

In this section we will introduce the results of texture analysis of aerial photographs carried out using a large-scale image-processing system (based on a 32-bit super minicomputer, DS 600/80, and a local parallel image processor, TOSPIX).

First, area partitioning of the aerial photographic image is carried out using light and shadow information for colors, and a number of characteristic quantities that reflect the texture of each area are extracted in the form of vectors with the same number of dimensions as the number of characteristic quantities. Next, various fuzzy clustering methods are used in the data set made from all of these vectors (there are only as many as there are partition areas), and we obtain the results of classifying the complete image into a number of fuzzy clusters. The final classification results contain subtle differences for each method used, and they are evaluated using entropy measures. We will explain fuzzy clustering methods, area partitioning that relies on the image and calculation of characteristic quantities, experimental results of texture classification, and evaluation using entropy, in that order.

The method of fuzzy clustering is an extension of clustering (normally called hard clustering), which is a versatile statistical method. Hard clustering involves introducing clear boundary lines and classifying the data set into a number of clusters, but it is often impossible to classify data that exist in a boundary area as completely belonging to any one cluster. Therefore, the degree of belonging is extended from two values $\{0, 1\}$ to $[0, 1]$ for discerning internal conditions, and the method in which the boundary data are formulated so as to permit attribution to more than one cluster is known as *fuzzy clustering.* Since it is an extremely natural way of thinking, many algorithms have been published and theoretically arranged.

Basically, if we let the data handled be $X = \{x_j\}_{j=1}^n (x_j \in R^s)$ and the number of clusters be c, the end result of fuzzy clustering can be expressed by one partition matrix U such that

$$U = [u_{ij}]_{i=1 \sim c, j=1 \sim n}. \tag{13.3}$$

Here u_{ij} is a numerical value from $[0, 1]$ and expresses the degree to which number j data x_j belongs to the ith cluster. However, there are two restrictions; we normalize so that the sum of degrees of attribution for each datum equals 1,

$$\sum_{i=1}^c u_{ij} = 1 \quad \text{for} \quad {}^\forall j = 1 \sim n, \tag{13.4}$$

and in addition, there is always a correct degree of attribution for each cluster:

$$\sum_{j=1}^n u_{ij} > 0 \quad \text{for} \quad {}^\forall i = 1 \sim c. \tag{13.5}$$

In other words, fuzzy clustering means finding a partition matrix U such that the restrictions in Equations (13.4) and (13.5) are satisfied. A number of algorithms for this have been published, but some are difficult to put into practical use. Of the standard algorithms published, the one that works well for the most part and can be used for large-scale data is the fuzzy C-means method, and the program for it is FUZZY ISODATA (an extension of the C-means method from hard clustering and ISODATA). This is an optimal partition classification method and uses the sum of internal squares as the homogeneity standard for the cluster. This is handled as a "weighted minimal square" problem, and it is a repetitive method in which an appropriate attribution function is devised as a weight.

We can consider

$$f = \sum_{i=1}^c \sum_{j=1}^n \sum_{k=1}^c g[w(x_i), u_{ij}] d(x_j, v_k) \tag{13.6}$$

as an objective function of the implementation index. Here $w(x_i)$ is the *a priori* weight for each datum, and $d(x_j, v_k)$ is the degree of dissimilarity of data x_j and supplemental element v_k (which can be thought of as the central vector for the kth cluster). The degree of dissimilarity is defined as a measure that satisfies these two axioms:

$$d(x_j, v_k) \geq 0 \tag{13.7}$$

$$d(x_j, v_k) = d(v_k, x_j), \tag{13.8}$$

and it is a concept that is weaker than distance. If we exchange v_k and U, we get

$$\min\{f \,|\, v_k \in R^s, U = [u_{ij}]: \text{Eq.}(13.4), \text{Eq.}(13.5)\}, \qquad (13.9)$$

and this is handled as a optimalization problem for f. When we use the objective function

$$f_p(U, \{x_j\}, \{v_k\}) = \sum_{i=1}^{c} \sum_{j=1}^{n} \sum_{k=1}^{c} (u_{ij})^p \| x_j - v_k \|^2 \qquad (1 \leq p < \infty), \quad (13.10)$$

one program for determining U by means of a repetitive method of local minimalization is FUZZY ISODATA, developed by Bezdek, et al. (see Fig. 13.6). The end result U varies depending on the value of parameter p in equation (13.10), and when $p = 2$, we know that we obtain results in which each cluster turns out to be a super ellipse.

There are several variations of FUZZY ISODATA, but here we bring up only three variations of the algorithm developed by Selim and Ismail, TN, TD, and TW.[4] They are all repetitive methods, but TN makes use of a threshold value for the number of clusters, TD makes use of one for $\| x - v_k \|^2$, and TW one for the value of u_{ij}. Here we exchanged three values for parameter p, $p = 1.2, 1.6, 2.0$ in TN, TD, TW, and FUZZY ISODATA, and we used a total of $12(= 4 \times 3)$ methods.

Next we will describe the method for constructing data set $\{x_j\}_{j=1}^{n}$, when characteristic extraction is carried out for the image. The image we are dealing with is an RGB color digital image and is constructed of 512×512 R(red), G(green) and B(blue) 8-bit($=256$ gradations) elements per frame. Therefore, we will consider the average light and dark value and standard deviation as characteristic quantities that reflect information for the complete image, and in addition bounded variation quantities (BVQ) for directions of $0°, 45°, 90°$, and $135°$ as local characteristic quantities that reflect the effects of information from one part on the whole. This is a total of six per frame, and we let one data vector x_j be expressed by an $s = 18(= 6 \times 3)$ dimensional vector. BVQ was proposed as local characteristic quantity using the concept of total variation of bounded variation functions for which there is a small number of calculations.[5] Actually, 18 characteristic quantities are calculated for each area partition, and the data set $\{x_j\}_{j=1}^{n}$ is constructed. The concrete procedure for the area partitions is carried out as follows. First, two threshold values are found for each R, G, and B component of the color image using light/shade histograms. Next, the image is converted into a three-value system by means of these two threshold values for each of the components. If there are many isolated points when this is done, isolated point exclusion (5×5 central value

```
C
C
C       FUZZY  ISODATA
C
C
        SUBROUTINE  ISO(E,N,IC,IS,P,L)
        REAL X(1024,18),U(10,1024),V(10,18),Y(18,1024),VA(1024,18)
        COMMON /XU/ X,U
C
C
      L=0
    5 DO 10 I=1,IC
        UPS=0.0
        DO 20 J=1,N
      UP=U(I,J)**P
          DO 30 II=1,IS
          VA(J,II)=X(J,II)*UP
   30     CONTINUE
        UPS=UPS+UP
   20   CONTINUE
        DO 40 II=1,IS
        VAS=0.0
          DO 50 J=1,N
          VAS=VAS+VA(J,II)
   50     CONTINUE
        IF(UPS.EQ.0.0) GO TO 40
        V(I,II)=VAS/UPS
   40   CONTINUE
   10 CONTINUE
        DO 60 I=1,IC
          DO 70 J=1,N
          UAN=0.0
            DO 80 II=1,IS
            UA=X(J,II)-V(I,II)
            UAN=UAN+UA*UA
   80       CONTINUE
          US=0.0
          DO 90 K=1,IC
            UDN=0.0
            DO 100 II=1,IS
            UD=X(J,II)-V(K,II)
            UDN=UDN+UD*UD
  100       CONTINUE
            IF(UDN.EQ.0.0) GO TO 90
            UDS=(UAN/UDN)**(1.0/(P-1.0))
            US=US+UDS
   90     CONTINUE
          Z=U(I,J)
          IF(US.EQ.0.0) GO TO 70
          U(I,J)=1.0/US
          Y(I,J)=ABS(U(I,J)-Z)
   70     CONTINUE
   60 CONTINUE
        L=L+1
        DO 110 I=1,IC
          DO 120 J=1,N
          IF(Y(I,J).GT.E) GOTO 5
  120     CONTINUE
  110 CONTINUE
        RETURN
        END
```

Fig. 13.6. Fortran-based FUZZY ISODATA

filter) is performed. In addition, the contour lines for each component are extracted, and a thin line process in which the three components are laid on top of each other is carried out.

Figure 13.7 shows a real example of this procedure (the colors have been converted to black and white). (a) is the RGB components of the original image (top left figure in (d)), and the three-value conversion of these is shown in (b). (c) shows the conversion after the extraction of contour lines. For each area of these partition results, the previously mentioned $18(=6 \times 3)$ characteristic quantities are calculated, and a set of 18 dimensional vectors is constructed (in this example the number of pieces of data n, that is, the number of areas partitioned in (c), is about 1500). For this data set, 12 methods of fuzzy

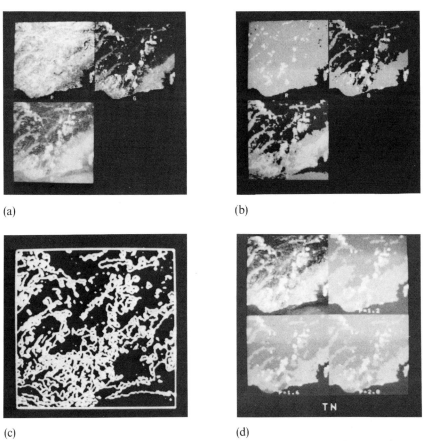

(a)   (b)

(c)   (d)

Fig. 13.7. Example of Results for Texture Analysis of Aerial Photographs

clustering are carried out (the number of clusters c is 5). From this, 12 partition matrices are obtained. Of these, the three for TN (corresponding to $p = 1.2, 1.6, 2.0$) are shown in (d) (each datum x_j is shown by the light/shade gradation [showing level 5($=c$)] corresponding to cluster i, which maximizes the value of u_{ij}).

The 12 results $\{U_k\}_{k=1}^{12}$ obtained in this way vary a little from each other. Entropy measures are used for comprehensive evaluation. First, the numerical values on $[0, 1]$ for each component are dispersed among the eleven values made up of tenths. In addition, we let the twelve methods be evaluated equally, and give each a weight of $1/12$ (frequency of appearance). The entropy value is calculated by means of the way in which the appearance of the components (i, j) of U_k is distributed in the 11 evaluation values, according to the differences in the methods, for each cluster $i(=1 \sim c)$ and each datum $(=1 \sim n)$. The averages for j in each i are given in Table 13.2(a). In the same way, the values of average entropies found for each pair of clusters and the reciprocal average entropies are given in Table 13.2(b).

From Table 13.2 we understand the following facts. First, the value for average entropy H for A_1 (ocean) is near zero and smaller than the other values. From this we know that the ocean part of the areas is clear. In contrast, $\bar{H}$ for the A_4 (valley) sections is large, and we know that the results vary with the method used. In addition, the reciprocal relation for A_3 (plains) and A_4 (valleys) shows that the boundary areas have been given classification evaluations that closely resemble each other. In this way it is possible to make a detailed texture analysis, using a comprehensive evaluation of the results obtained, which makes use of several comparatively good algorithms.

Table 13.2.   Average Entropy Values for Results in Fig. 13.7

(a)

$\bar{H}(X, A_1)$	0.163	A_1: ocean
$\bar{H}(X, A_2)$	1.895	A_2: mountains
$\bar{H}(X, A_3)$	1.853	A_3: plains
$\bar{H}(X, A_4)$	2.008	A_4: valleys
$\bar{H}(X, A_5)$	0.746	A_5: clouds

(b)

	$\bar{H}$	$\bar{I}$
X, A_1, A_2	0.379	1.679
X, A_1, A_3	0.913	1.103
X, A_1, A_4	0.605	1.566
X, A_1, A_5	0.199	0.709
X, A_2, A_3	0.377	3.372
X, A_3, A_4	0.268	3.635
X, A_3, A_5	0.401	2.240
X, A_3, A_4	0.162	3.700
X, A_3, A_5	0.722	1.877
X, A_4, A_5	0.362	2.392

REFERENCES

(1) Hirota, K., *Image Pattern Recognition*, McGraw-Hill Books, Tokyo (1984).

(2) Hirota, K., *Intelligent Robots with Fuzzy Control*, Chapter 3, McGraw-Hill Books, Tokyo (1985) (in Japanese).

(3) Bezdek, J. C., *Pattern Recognition with Fuzzy Objective Function Algorithms*, Plenum Press, New York (1981).

(4) Selim, S. Z., and Ismail, M. A., "Soft Clustering of Multidimensional Data: A Semi-Fuzzy Approach," *Pattern Recognition*, **17**, 5, pp. 559–568 (1984).

(5) Hirota, K., "The Bounded Variation Quantity (B. V. Q.) and Its Applications to Feature Extractions," *Pattern Recognition*, **15**, 2, 93–101 (1982).

Chapter 14

DATABASES

For efficient processing of data, the data must be stored and retrieved uniformly and then processed. Because of these requirements, research on databases began in the 1960's, and E. F. Codd's paper on a relational model of data provided an opportunity for extremely enthusiastic research in the 1970's.

In fields that directly involve people, such as human-machine systems, natural language processing, and decision making, there is a great deal of ambiguous data that cannot or need not be precisely defined.

If we try to use this ambiguous data constructively, it becomes extremely complicated to manipulate within the framework of the databases proposed up to now. Therefore several databases based on fuzzy theory have been proposed. We will describe this type of database using many examples.

14.1 STANDARD DATABASES

Before we talk about fuzzy databases, let us give a simple summary of standard databases. Most standard databases are classified according to a model of how the data is viewed by the user. A great number of data models have been proposed, but the three that have been put to practical use are tree-structure

217

(hierarchical) models, network models, and relational models. These have been studied enthusiastically since E. F. Codd proposed his relational model in 1970,[1] and in the 1980's practical systems have been released.

Since most fuzzy databases are based on relational models, we will just give a simple description of them.

In a relational model, the database is a group of relations. The relations are essentially the same as the relations of set theory, and are usually expressed in the form of tables as in Fig. 14.1, which shows a database made up of three relations: PART, SUPPLIER, and SP (shipping). In the tables, the data in each row are connected; for example, if we look at the first row of the PART relation, it shows us that part number (P#) is P1, the part name (PNAME) nut, the color red, the weight 12, and the storage location (CITY) London. The lineup of one row is called a tuple, and here it is written using brackets $\langle \ \rangle$, as in $\langle$P1, nut, red, 12, London$\rangle$. The names given to the columns, such as P# and PNAME are called *attributes*. The elements of the table, such as P1, nut, etc., are called *attribute values*. The set of attribute values of an attribute is called the *domain of the attribute*. For example, the domain of the

PART

P#	PNAME	COLOR	WEIGHT	CITY
P1	nut	red	12	London
P2	bolt	green	17	Paris
P3	screw	blue	17	Rome
P4	screw	red	14	London

SUPPLIER

S#	SNAME	STATUS	CITY
S1	Smith	20	London
S2	Jones	10	Paris
S3	Blake	30	Paris

SP

S#	P#	QTY
S1	P1	3000
S1	P2	2000
S1	P3	4000
S2	P1	3000
S2	P2	4000
S3	P2	2000

Fig. 14.1. Tabular Expression of Database

attribute P# for the PART relation is all of the possible part numbers for
P# in the database.

If we write down just the framework of the database in Fig. 14.1, we get

PART(P#, PNAME, COLOR, WEIGHT, CITY)

SUPPLIER(S#, SNAME, CITY)

SP(S#, P#, QTY).

There are advantages to defining the database as a group of relations, as
follows:

(1) simple, easily understood data model
(2) high-level data retrieval and manipulation languages based on set
     operations (relational algebra and relational calculus)
(3) logical design and relational semantics for the database, based on
     normalization theory
(4) theoretical handling of consistency and completeness of data.

For details of these, see References (2) and (3).

However, there is discussion as to whether the relational model is good
enough to express the complicated data that exist in the real world, and new
concepts for the databases have recently been proposed.

14.2 FUZZY DATABASES

In order to store, retrieve, and process the ambiguous data that exist in
the real world, data models used up to now have not been good enough, so
fuzzy databases have been proposed. As far as the authors can determine,
the term "fuzzy database" was first used in Kunii's DATAPLAN.[4] Almost
all fuzzy databases are extensions of relational models, but the formulation
differs according to the type of ambiguity they are intended to express and
manipulate.

14.2.1 Ambiguous Queries

First, we can consider a data model for standard relations, which uses fuzzy
theory in the retrieval conditions for queries.

V. Tahani describes a processing method[5] for fuzzy queries based on a data
manipulation language called SEQUEL[6]. Let us consider an employee

EMP

NAME	AGE	SALARY	EMP-YEAR
Anderson	30	20 000	1974
Brown	30	15 000	1974
Long	25	40 000	1972
Nelson	55	20 000	1950
Smith	25	25 000	1975

Fig. 14.2. Employee Database

database like that in Fig. 14.2. This is a simplification and slight modification of an example in Reference (5). Let us first consider the set of fuzzy linguistic values that can be used for the attributes. For example, we can use

T(AGE) = {old, young, very old, not old, more or less old,...}

T(SALARY) = {high, low, very high, more or less high,...}

T(EMP-YEAR) = {recent, more or less recent, very recent,...},

where we do not ask fuzzy queries about names. By doing so, we are able to write the following queries, which contain fuzzy conditions.

Query 1

Of people who are young or recently employed, what are the names of those who have high salaries?

> SELECT   NAME
> FROM   EMP
> WITH   (AGE = "young" or EMP-YEAR = "recent")
>   and SALARY = "high"

Query 2

Of people who are old, or who are employed more or less recently and with very high salaries?

> SELECT NAME, AGE, SALARY
> FROM   EMP
> WITH   AGE = "old" or
>   EMP-YEAR = "more or less recent"
>   and SALARY = "very high"

Let us consider the method for processing Query 1. The first tuple in the EMP database, $\langle$Anderson, 30, 20000, 1974$\rangle$, has a value of $\mu_{young}(30)$ for the membership function of "young" for the AGE attribute value of 30. Since "young" is in T(AGE), the method for calculating $\mu_{young}(30)$ is given, say, $\mu_{young}(30) = 0.5$. In the same way, the EMP-YEAR is 1974, so we let $\mu_{recent}(1974) = 0.6$. The salary is 20,000, and we let $\mu_{high}(20,000) = 0.5$. In this case, the definition of the membership function of "recent" conforms to 1977, the year Reference (5) was published.

Since the "and" and "or" in the condition are defined as

$$\gamma(p \text{ and } q, t) = \min(\gamma(p, t), \gamma(q, t))$$

$$\gamma(p \text{ or } q, t) = \max(\gamma(p, t), \gamma(q, t)),$$

respectively, the whole query has a degree of

$$\min(\max(0.5, 0.6), 0.5) = \min(0.6, 0.5) = 0.5.$$

This can be carried out for the remaining tuples. The results are summarized in Table 4.1. The necessary domain for the tuples that satisfy the query is selected and we have the following fuzzy relation (fuzzy set):

$$0.5/\text{Anderson} + 1/\text{Long} + 0.8/\text{Smith}.$$

Besides the above method, modal logic,[7] intuitive logic,[8] and denotational semantics[9] are used in the interpretation of queries. In addition, there have been proposals for various retrieval methods in the field of information retrieval (document and fact retrieval).

14.2.2 Extension of Data Models

In order to express ambiguous data, we must extend data models. The simplest extension is making fuzzy relations by adding grades to standard relations. The databases in L. A. Zadeh's PRUF[10] and J. F. Baldwin's FRIL[11] are of this form. However, we will not cover these in this book, because the objects of these studies are not the databases themselves.

Table 14.1. Truth Values for Fuzzy Queries

Condition / Tuple	AGE = "Young"	EMP-YEAR = "Recent"	SALARY = "High"	Complete Query
⟨Anderson, 30, 20 000, 1974⟩	0.5	0.6	0.5	0.5
⟨Brown, 30, 15 000, 1974⟩	0.5	0.6	0	0
⟨Long, 25, 40 000, 1972⟩	1	0	1	1
⟨Nelson, 55, 20 000, 1950⟩	0	0	0.5	0
⟨Smith, 25, 25 000, 1975⟩	1	0.8	0.8	0.8

Buckles and Petry's Fuzzy Database

Buckles and Petry extended a relational model;[12-14] introduced similarity relations,[15] which are an introduction of fuzziness into equivalence relations; and extended operations for relational models known as relational algebra.

Their database is made up of three parts: domain sets, similarity relations for domains, and relations. An example is shown in Fig. 14.3, and this is a slight modification of an example in Reference (12).

Let us consider the execution of

$$R \leftarrow (\text{project}(R_1 \text{ over } A) \text{ with LEVEL}(A) = 0.4)$$

for this type of database. "Project" is a relational algebra operation for picking the data for the indicated domain only. Since this calculation says that relation R_1 is projected on domain A, we get $\{a_1, a_3, a_5, a_1\}$. And since there are two a_1's, we get $\{a_1, a_3, a_5\}$. However, we complete this based on similarity relations, instead of equivalence relations. What follows "with" indicates the threshold value, meaning that we view elements with a similarity of 0.4 or more for domain A as being the same. If we take a look at the similarity relations for Fig. 14.3, the similarity of $\langle a_1, a_1 \rangle$ is 1, that of $\langle a_1, a_3 \rangle$ 0.3, $\langle a_1, a_5 \rangle$ 0.5, and $\langle a_3, a_5 \rangle$ 0.3, so when we sum up a_1 and a_5, we get $\{\{a_1, a_5\}, a_3\}$ as shown in Fig. 14.4.

This result includes a group of more than one element in the relation. Since this group cannot be expressed by a standard relational model, their data

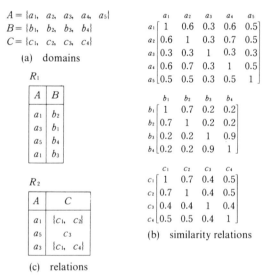

$A = \{a_1, a_2, a_3, a_4, a_5\}$
$B = \{b_1, b_2, b_3, b_4\}$
$C = \{c_1, c_2, c_3, c_4\}$

(a)   domains

R_1

A	B
a_1	b_2
a_3	b_1
a_5	b_4
a_1	b_3

R_2

A	C
a_1	$\{c_1, c_2\}$
a_5	c_3
a_3	$\{c_1, c_4\}$

(c)   relations

	a_1	a_2	a_3	a_4	a_5
a_1	1	0.6	0.3	0.6	0.5
a_2	0.6	1	0.3	0.7	0.5
a_3	0.3	0.3	1	0.3	0.3
a_4	0.6	0.7	0.3	1	0.5
a_5	0.5	0.5	0.3	0.5	1

	b_1	b_2	b_3	b_4
b_1	1	0.7	0.2	0.2
b_2	0.7	1	0.2	0.2
b_3	0.2	0.2	1	0.9
b_4	0.2	0.2	0.9	1

	c_1	c_2	c_3	c_4
c_1	1	0.7	0.4	0.5
c_2	0.7	1	0.4	0.5
c_3	0.4	0.4	1	0.4
c_4	0.5	0.5	0.4	1

(b)   similarity relations

Fig. 14.3.   Example of Fuzzy Database by Buckles and Petry

Fig. 14.4. Relation R

model has been extended so that it can express this kind of relation. Relation R_2 in Fig. 14.3 also includes a group of two elements.

In addition, operations for a relational model, which are called *relational calculus*, are extended in this database.

Fuzzy Database by Umano et al.

Buckles and Petry's fuzzy database introduces fuzziness by means of similarity relations. On the other hand, Umano, *et al.*, express the ambiguity of the data itself by means of possibility distributions,[16] and they have proposed a possibility distribution-relational model, with possibility distributions as attribute values, based on which data manipulation language is designed and implemented.[17] In Reference (5), Tahani also comments on ambiguity in the data themselves, but he does not touch on methods for processing queries in such cases.

We can write relations for this data model, as shown in Fig. 14.5.[18] Since most of what can be expressed by this data model is in this relation, let us look at the meaning in some detail. In the first row, 23-year-old Tom has a child named Ted. This is unambiguous data. If we look at the second and third rows, we see that Susan is 35 and has two children, John and Mark. Next is Richard, who is 40 and has one child, whose name is Judy or Anna. This attribute value $\{Judy, Anna\}_p$ expresses a possibility distribution; this does not mean that it has both values, Judy and Anna, but rather that it is one or the other, and we cannot be sure of which from current information. If we have

PERSON

NAME	AGE	CHILD NAME
Tom	23	Ted
Susan	35	John
Susan	35	Mike
Richard	40	$\{Judy, Anna\}_p$
Raymond	young	unknown
Victor	unknown	undefined
Smith	$\{50, 51\}_p$	null

Fig. 14.5. Relation Including Possibility Distributions

$\{1/\text{Judy}, 0.7/\text{Anna}\}_p$, the possibility of its being Judy is greater than that of its being Anna. If he had two children, Judy and Anna, it must be divided into two rows as with Susan.

Next, Raymond's age is expressed by the possibility distribution "young"; he has one child, but the name is unknown. The possibility distribution "young" can be defined, for example, as

$$\text{young} = \{0.3/15, 0.6/16, 0.8/17, 1/18, 1/19, 1/20, 1/21,$$
$$1/22, 1/23, 0.9/24, 0.8/25, 0.7/26, 0.5/27,$$
$$0.3/28, 0.1/29\}_p.$$

"Unknown" can also be interpreted as a possibility distribution; since we do not know the name, we can say that it could be any name. Therefore, it is defined as the possibility distribution in which all names have a possibility of 1. (In the implementation this is handled specially.) The sixth row shows that Victor's age is unknown, and he has no children. This "unknown" is the possibility distribution in which all ages have a possibility of 1. In addition, "undefined" can be interpreted as a possibility distribution; since he has no children, we can say that there is no possibility of any name, so it is defined as the possibility distribution in which all names have a possibility of zero. (This also requires special handling in the implementation.)

This kind of data could not be expressed in the databases used up to now, so it was omitted. However, it is necessary for distinguishing people who have not even been examined as to whether they have children or not.

Finally, Smith is either 50 or 51 years old, and his child's name is "null." This "null" is a special value; it was examined, but we do not even know whether it is "unknown" or "undefined." In other words, it shows that we do not know whether Smith has any children or not. (This also requires special handling in the implementation.)

Using this kind of data model, we can express the ambiguity of the attribute values themselves. Let us see what kind of retrieval are made with this kind of database.

Let us consider a nonfuzzy query such as "Find people with ages over 25." There are data that clearly satisfy the condition and data that clearly do not satisfy it. In addition, attribute values can be possibility distributions, so there are those that possibly satisfy it. In fact, Tom clearly does not satisfy the condition, and Susan clearly satisfies it. If we look at the possibility distribution "young" for Raymond, it is possible that he is over 25 (values exist for which the possibility is nonzero), so there is a possibility that the conditions are satisfied. Victor's age is "unknown," so any age is possible; therefore, there is a possibility that Victor satisfies the condition. As for Smith, he is 50 or 51, so either way he clearly satisfies the condition. If we have a person who is "old," all ages for which the possibility of "old" is nonzero satisfy the condition, so he clearly satisfies the condition. Note that those that possibly satisfy the condition satisfy it with some elements whose possibility is nonzero, but since others whose possibility is nonzero do not satisfy the condition, it is possible that the former group do not satisfy it. However, it is more constructive to employ the possibility of satisfying here.

Summing up the above, the results can be divided into the following three sets.

clearly satisfy   = 1/Susan + 1/Richard + 1/Smith

possibly satisfy   = 1/Raymond + 1/Victor

clearly don't satisfy = 1/Tom

Since the query is "25 years old or older?," which is not fuzzy, the grade of membership for each is either zero or 1 (those with zero are not written down).

Reference (17) shows methods for processing queries with fuzzy conditions such as "Find the names and ages of people about 25 years old" and "Find the names and ages of people who are young." In these cases, the membership grades for each of the above sets takes a value between zero and 1.

Furthermore, data manipulation language for carrying out this kind of processing has been designed and implemented.[17] In the following we will define the relation for PERSON of Fig. 14.5 and discuss retrieval methods.

First, let us define the relation PERSON. For this we use DEFR (define relation).

DEFR PERSON ⟨NAME: CHAR, AGE: INTEGER,
CHILD_NAME: CHAR⟩ DEFEND

where CHAR means a character string.

Second, we define the possibility distributions for the attributes. In the present version, possibility distributions must be given names and the names must begin with $.

$YOUNG = FSET(0.3/15, 0.6/16, 0.8/17, 1/18, 1/19, 1/20, 1/21, 1/22, 1/23, 0.9/24, 0.8/25, 0.7/26, 0.5/27, 0.3/28, 0.1/29);

$50-51 = FSET(1/50, 1/51);

$JUDY-ANNA = FSET(1/JUDY, 1/ANNA);

where FSET is an operator for constructing a fuzzy set, and possibility distributions are expressed by fuzzy sets within the system. In addition, "unknown," "undefined," and "null" are written $UNKNOWN, $UNDEFINED, and $NULL, respectively.

Third, we add the data to the relation PERSON.

INSERT PERSON ⟨TOM, 23, TED⟩, ⟨SUSAN, 35, JOHN⟩
⟨SUSAN, 35, MIKE⟩
⟨RICHARD, 40, $JUDY–ANNA⟩
⟨RAYMIND, $YOUNG, $UNKNOWN⟩,
⟨VICTOR, $UNKNOWN, $UNDEFFINED⟩,
⟨SMITH, $50–$51, $NULL⟩ IEND

Now, the relation PERSON in Fig. 14.5 has been constructed.

We can write the query "Find people who are 25 years old or older" as follows:

QUERY A(NAME = X):
PERSON(NAME = ?X, AGE = ?Y);
AGE (*Y, 25);
QEND

This says that those that satisfy all of the conditions between QUERY and QEND will be put into a set (generally a relation) named A. The resulting set A has the attribute NAME, and the values for this attribute are obtained from variable X. Variable X in line 2 is defined as the attribute NAME of relation PERSON and Y as the attribute AGE. The ? shows that the variable is defined here and that the value for the specified attribute is to be assigned. The third line selects values for Y equal to 25 or more. The asterisk (*) shows that a previously defined variable is used. In other words, this query means that the values for the attribute NAME of the relation PERSON for which the AGE attribute values are 25 or more are collected in a set named A. The results that clearly satisfy the condition are put into A@1, and those that possibly satisfy them into A@2. Those that clearly do not satisfy are discarded. The results are as follows:

A@1 = FSET(1/SUSAN, 1/RICHARD, 1/SMITH);

A@2 = FSET(1/RAYMOND, 1/VICTOR);

One method for the queries "Find the names of people who are about 25 years old" and "Find the names of people who are young" involves defining possibility distributions called ABOUT_25 and YOUNG, substituting

FEQ(*X, @ABOUT_25); or FEQ(*X, @YOUNG); for line 3 in the above query and retrieving by means of equality of the possibility distributions. Another involves using the DEFP (define fuzzy predicate) operator for ABOUT_25 and YOUNG, defining the fuzzy predicates, and making line 3 of the query ABOUT_25(*X); or YOUNG (*X); (the two methods differ in processing).

We discussed a relational model that has possibility distributions for attribute values and its manipulation language. This data manipulation language is implemented in the FSTDS system, a Fortran-based fuzzy set manipulation system.[18]

Other Fuzzy Databases

Besides ambiguity in the values of the data themselves discussed above, data can have other types of ambiguity, such as ambiguity in the associations among them. In order to express this type of data, Umano, *et al.*, have further extended their data model and have proposed a possibility distribution–fuzzy relational model. Almost all parts of the relational algebra[19] and relational calculus[20] have been formulated for this kind of data model.

H. Prade and C. Testemale have somewhat extended the possibility distribution–relational model discussed in this section.[21] Possibility distributions for the attributes can be written in the same way, but they have added a special value when the attributes cannot be applied (for example, when the attribute values do not exist). Furthermore, the relational algebra operations are defined using possibility and necessity measures. The results are obtained as two fuzzy sets based on possibility and necessity measures. The data manipulation language is implemented in MacLisp.

M. Zemankova-Leech and A. Kandel have proposed a possibility distribution–relational model database.[22,23] This database is made up of three parts; (1) value database (VDB), (2) explanatory database (EDB), and (3) conversion rules. The value database is essentially the same as the possibility distribution–relational model in this section (there are a few more data types). The explanatory database is the collection of definitions of named fuzzy sets and fuzzy relations, and can be different for each user. The conversion rules are for processing modifiers and qualifiers. The data manipulation language is implemented in RIM (Relational Information Management System), which is based on relational algebra and was developed by Boeing.

In this chapter, we have explained several databases that are based on fuzzy theory. As can be seen from the above discussion, theoretical handling and implementation are comparatively easy if the data model is established on a level near that of a standard relational model, but the ability to express data from the real world is lost. On the other hand, if we establish a data model geared to express of real-world data, theoretical handling and implementation become difficult. A data model that harmonizes these two objectives needs to be found, but at present research into fuzzy databases has just started, and we are at the stage in which various data models are being proposed. Since fuzzy databases will play essential roles in decision support systems and expert systems, there are great expectations for future developments.

REFERENCES

(1)   Codd, E. F., "A Relational Model for Large Shared Data Banks," *Communications of the ACM*, **13**, pp. 377–387 (1970).

(2)   Uemura, S., *Foundation of Database Systems*, Ohmsha, Tokyo, Japan (1979) (in Japanese).

(3)   Date, C. J., *An Introduction to Database Systems*, Vol. 1 — Third Edition, Addison Wesley, Reading, Mass. (1981).

(4)   Kunii, T. L., "DATAPLAN: An Interface Generator for Database Semantics," *Information Sciences*, **10**, pp. 279–298 (1976).

(5)   Tahani, V., "A Conceptual Framework for Fuzzy Query Processing—A Step toward Intelligent Database Systems," *Information Processing & Management*, **13**, pp. 289–303 (1977).

(6)   Chamberlin, D. D., and Boyce, R. F., "SEQUEL: A Structured English Query Language," Proceedings of ACM–SIGFIDET Workshop, Ann Arbor (May 1974).

(7)   Lipski, W., "On Data Bases with Incomplete Information," *Journal of the ACM*, **28**, pp. 41–70 (1981).

(8)   Jagermann, M., "Information Storage and Retrieval Systems with Incomplete Information," *Fundamenta Informaticae*, **2**, pp. 17–41 (1978).

(9)   Vassiliou, Y., "Null Values in Database Management—A Denotational Semantics Approach," Proceedings of ACM-SIGMOD 1979 International Conference on Management of Data, Boston, Mass., pp. 162–169 (1979).

(10) Zadeh, L. A., "PRUF—A Meaning Representation Language for Natural Languages," *International Journal of Man-Machine Studies*, **10**, pp. 395–460 (1978).

(11) Baldwin, J. F., "FRIL—A Fuzzy Relational Inference Language," *Fuzzy Sets and Systems*, **14**, pp. 155–174 (1984).

(12) Buckles, B., and Petry, F., "A Fuzzy Model for Relational Databases, *Fuzzy Sets and Systems*," **33**, pp. 213–226 (1982).

(13) Buckles, B., and Petry, F., "Fuzzy Databases and Their Applications," in Gupta,

M. M., and Sanches E., eds., *Fuzzy Information and Decision Processes*, pp. 361–371, North-Holland, Amsterdam (1982).

(14) Buckles, B., and Petry, F., "Query Languages for Fuzzy Databases," in Kacprzyk, J., and Yager, R. R., eds., *Management Decision Support Systems Using Fuzzy Sets and Possibility Theory*, pp. 241–252, Verlag TÜV Reinland, Köln, Germany (1985).

(15) Zadeh, L. A., "Similarity Relations and Fuzzy Orderings," *Information Sciences*, **3**, pp. 177–200 (1971).

(16) Zadeh, L. A., "Fuzzy Sets as a Basis for a Theory of Possibility," *Fuzzy Sets and Systems*, **1**, pp. 3–28 (1978).

(17) Umano, M., "FREEDOM-0: A Fuzzy Database System," in Gupta, M. M., and Sanchez, E., eds., *Fuzzy Information and Decision Processes*, pp. 339–347, North-Holland (1982).

(18) Umano, M., Mizumoto, M., and Tanaka, K., "FSTDS System: A Fuzzy Set Manipulation System," *Information Sciences*, **14**, pp. 115–159 (1978).

(19) Umano, M., "Retrieval from Fuzzy Database by Fuzzy Relational Algebra," Proceedings of IFAC Symposium on Fuzzy Information, Knowledge Representation and Decision Analysis, Marseille, France, pp. 1–6 (1983).

(20) Umano, M., Fukaumi, S., Mizumoto, M., and Tanaka, K., "On Retrieval Processing from Fuzzy Databases," Preprints of Working Group of IEICE of Japan, **80**, 204, pp. 45–54, AL 80-50 (in Japanese).

(21) Prade, H., and Testemale, C., "Generalizing Database Relational Algebra for the Treatment of Incomplete or Uncertain Information and Vague Queries," *Information Sciences*, **34**, pp. 115–143 (1984).

(22) Zemankova-Leech, M., and Kandel, A., *Fuzzy Relational Data Bases—A Key to Expert Systems*, Verlag TÜV Reinland, Köln, Germany (1984).

(23) Zemankova-Leech, M., and Kandel, A., "Implementing Imprecision in Information Systems," *Information Sciences*, **37**, pp. 107–141 (1985).

Chapter 15

INFORMATION RETRIEVAL

When we are collecting a certain kind of information, what we want exists in our minds only as an ambiguous conceptual arrangement. In addition, it is extremely difficult for us to state clearly what we need to confirm that concept. For example, let us consider the case in which we collect the information necessary for our own research by searching for documents. We judge and obtain the documents and books that have the content we require by information such as author, title, and keywords; we scrutinize the contents, eliminate what is unnecessary, and continue searching to supplement the parts for which there is not enough information. If at this time there is a librarian that knows the details of the collection, the efficiency of our search is greatly increased. With some conceptual understanding of the information that the researcher needs, the librarian can consider the range of the collection for himself or herself and give appropriate advice. The initial development of technology related to the recording of information has been accomplished in the form of databases, which are one form of computer technology. In these, information storage must be organized and accumulated based on the assumption of a large number of unspecified users. The introduction of an intelligent interface between the computer and user that would take the place of the above-mentioned librarian, creating a highly efficient man-machine interaction, is indispensable for solving problems, such as system planning,

organization, and decision making, all of which require dialogue with the
large amount of information in the database. Here we will describe the authors' intelligent interface, which uses fuzzy sets, for information retrieval.[1]

15.1 INFORMATION RETRIEVAL AND MODELING OF ESTIMATION PROCESSES USING FUZZIFICATION FUNCTIONS

Information retrieval can be considered a process of making connections
among most existing concepts and of structuring comprehensive knowledge.
For example, in a document search, the documents and books required by the
researcher can be considered existing concepts for the production of requested
information, that is, comprehensive knowledge. For expression of concepts
within the computer, there are systems such as a semantic network based on
Quillian's Semantic Memory[2] and Minsky's frame,[3] but all of these consider concepts as nodes in graphs or "frames," and the group of attributes that
describe the concept is recorded in nodes, on branches joining these nodes, or
in "slots" within the frame. When we consider documents to be the existing
concepts, the group of keywords can be seen as a group of attributes that
describes that concept—that is, the contents of the documents. In standard
document retrieval using keywords, for example, if the researcher inputs keyword x as in Fig. 15.1, the output is documents A, B,..., which have x as an
attribute. Which documents in the output are closest to the researcher's
requirements can be determined by finding out which have more of the
keywords that include the researcher's request concepts as attributes. In order

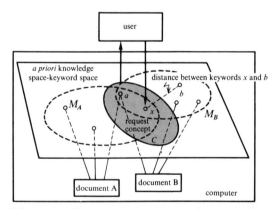

Fig. 15.1. Information (Document) Retrieval and *A Priori* Knowledge Space

to give the computer a process for inferring the researcher's request concepts from the single piece of information x given by the researcher, a structure that suggests the conceptual distance between keywords must be given to the keyword space in Fig. 15.1. For example, if the distance $l(x, y)$ that reflects the affinity based on a conceptual description of keywords x and y is defined and information like $l(x, a) < l(x, b)$ (keyword a has greater affinity with keyword x than keyword b does) can be used, the computer can recognize the fact that the researcher's request concept is included in x and a more than in x and b from the single piece of input information x. Based on this recognition of the request concept, the system gives documents that have a relevant range and common range of more attributes. In other words, in order to have the computer take the place of the experienced librarian mentioned above, that is, in order efficiently to coordinate the contents of the database with the researcher's request concepts, creating a structure based on the conceptual affinity of each attribute (keyword) of the keyword space is an indispensable goal. Here we will call structured keyword spaces *a priori* knowledge spaces.

15.1.1 *A Priori* Knowledge Spaces

When we think of documents as established concepts, the pair of keywords common to one document in some way carry connections with those concepts. We can say that the appearance of a larger number of documents means the conceptual description has a larger affinity. L. B. Doyle has quantified affinity, which corresponds to the similarity measure in psychology,[4–6] using a frequency value for the simultaneous appearance of keyword pairs, as shown in Fig. 15.2.[7] Here we will use the inverse $l(x, y)$ of the relative value for frequency of simultaneous appearance of keyword pair (x, y) in the same document to define their distance, and this frequency is given by the weight of

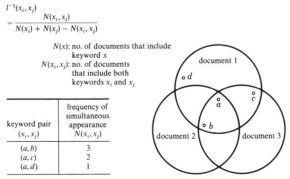

$$l^{-1}(x_i, x_j) = \frac{N(x_i, x_j)}{N(x_i) + N(x_j) - N(x_i, x_j)}$$

$N(x)$: no. of documents that include keyword x

$N(x_i, x_j)$: no. of documents that include both keywords x_i and x_j

keyword pair (x_i, x_j)	frequency of simultaneous appearance $N(x_i, x_j)$
(a, b)	3
(a, c)	2
(a, d)	1

Fig. 15.2. L. B. Doyle's Association Factor $l^{-1}(x_i, x_j)$

the link between keywords in Fig. 15.1. Using this weighted graph for *a priori* knowledge space X, it is possible for the computer efficiently to recognize the request concept that probably extends to several documents, by means of dialogue with the researcher.

15.1.2   Recognition of Requested Concepts for Retrieval Using Dialogue (Expression of Recognition Concepts as Fuzzy Sets in *A Priori* Knowledge Spaces)

As mentioned before, it is difficult for the researcher to state completely and clearly the information he requires. However, when a particular attribute (keyword) is exhibited, he can easily say whether it describes the concept he is requesting. We now let keyword x_s be known to be an attribute that describes request concept C (that is, an element of ordinary set C). At this point, besides having the direct information $x_s \in C$, the computer can infer the degree of inclusion in the request concept for another keyword x in the space, using the distance in the *a priori* knowledge space. By defining fuzzification function $f_{xs}(x)$ shown in Fig. 15.3(a), we let the keyword shown in Fig. 15.4(a) be the

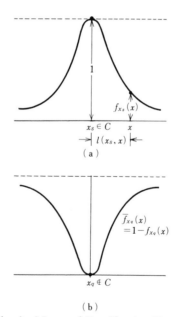

Fig. 15.3.   Modeling by Means of Fuzzification Function for Estimation Effectiveness

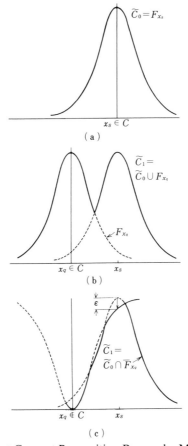

Fig. 15.4. Request Concept Recognition Process by Means of Estimation

support set in the computer's *a priori* knowledge space in order to give the degree of inclusion quantitatively, and fuzzy set F_{x_s} is formed. We can think of this F_{x_s} as the computer's first recognition $\tilde{C}_0 = F_{x_s}$ for the acquired concept request based on *a priori* knowledge used with the sole piece of input information "x_s belongs to request concept C." We let the information that x_q is also included in the request concept be obtained following x_s. At this point the computer's recognition is changed to $\tilde{C}_1 = \tilde{C}_0 \cup F_{x_q}$, as shown in Fig. 15.4(b). When the information that x_q is not included in the request concept is obtained, using the fuzzification function $\bar{f}_{x_q}(x)(=1 - f_{x_q}(x))$, shown in Fig. 15.3(b), which quantifies the decrease in the degree of inclusion in the request concept, we change the recognition concept to $\tilde{C}_1 = \tilde{C}_0 \cap \bar{F}_{x_q}(\neg \tilde{C}_1) = (\neg \tilde{C}_0) \cup (\neg \bar{F}_{x_q})$, as shown in Fig. 15.4(c).

As is clear from the above, inputting an attribute x_q which is included in the researcher's request concept, expands the range $\tilde{C}$ of the computer's concept recognition, which is set up as a fuzzy set on the *a priori* knowledge space. Inputting an attribute that is not included serves to limit range $\tilde{C}$. This process can be thought of as a modeling of the process in which the computer estimates the information to be included in the request concept from direct input information x_s and x_q. Using this function, the computer can achieve recognition of the concept the researcher requires as fuzzy set $\tilde{C}$, which is as close as possible to that concept, from the fragmentary information of the input.

15.1.3   Choice of Query Attributes Based on Recognition Concepts

The query attributes (keywords) that revise the computer's recognition $\tilde{C}$ of the request concept are chosen by the computer, with the exception of the first specification $x_s \in C$, which is made by the person searching for information. In this case the computer asks questions so that the person can clarify the ambiguous part as quickly as possible, and using the answers, the computer chooses attributes that reduce the ambiguity of the recognition of the request concept $\tilde{C}$ as much as possible. As an index of the ambiguity of fuzzy set $\tilde{C}$, we have Kaufmann's index of fuzziness I.[8] I, as is shown in Fig. 15.5, is given by the value of the square error when we consider $\tilde{C}$ and the smallest ordinary set $\bar{C}$ for the square error. In this case, $d(x)$ in the same figure can be thought of as a measure of the ambiguity of the inclusion of attribute x in $\tilde{C}$. Now we let the computer's recognition at the end of the $k-1$th recognition renewal be $\tilde{C}_{k-1}$. The computer chooses the value with the smallest average for the I values, $I(k)^+$ and $I(k)^-$, for $\tilde{C}_{k-1} \cup F_{x_k}$ and $\tilde{C}_{k-1} \cap F_{x_k}$ that is the value that

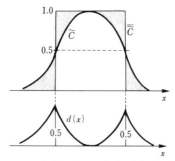

Fig. 15.5.   A Kaufmann's Ambiguity Index

minimizes the expected value for the ambiguity $I(k)$ of the newly formed recognition $\tilde{C}_k$. This calculation is done only for keywords other than those already used which have degrees of attribution to $\tilde{C}_{k-1}$ above a certain value. This kind of x_k exists in the neighborhood that has a maximum inclusion ambiguity $d(x_k)$ of (0.5), that is, in the border neighborhood of fuzzy set $\tilde{C}_{k-1}$.

15.1.4   Adaptive Change of the Estimation Strength in the Recognition Concept Revision Process

As shown in Fig. 15.6, x_i is a new query keyword and x_j is the keyword used in the previous query. Let us say that the answers given by the user for both of them are reciprocal. When the distance between x_i and x_j in the *a priori* knowledge space is small, the inference is that the boundary of request concept C exists between x_i and x_j. This is very important information for recognition of the request concept, and in order to preserve this information, it is desirable to prevent the effect of the estimate using input x_i from reaching the neighborhood of x_j. This operation is accomplished by an appropriate adaptive change of the width of the fuzzification function, using this information, as shown in the figure.

15.1.5   Termination of Recognition Concept Revision and Output of Results

When the number of attributes (keywords) in the *a priori* knowledge space in Fig. 15.1 is finite, request concept C can be completely determined by making queries about all of them. In other words, fuzzy set $\tilde{C}$, which expresses the recognition concept, can be made the same as C. However, with actual problems, revision is terminated after a certain degree of recognition is attained, that is, after the index of ambiguity $I(k)$ in the recognition concept has been reduced to a certain value. In this case, the following problem arises.

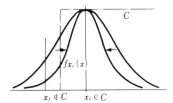

Fig. 15.6. Adaptive Change of Strength of Estimation in the Request Concept Border Area

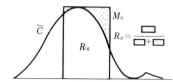

Fig. 15.7. Degree of Inclusion R_s of Existing Concept (Document) M_s for Recognition Concept $\tilde{C}$

When the recognition concept is wide in a vague way there are naturally cases when we can think of C as being a disconnected group of sets. If the estimation strength is small in this kind of case, $\tilde{C}$ can be close enough to one part of C so that the $I(k)$ value falls below the set value, and recognition revision is terminated. We can define the Euclidean distance $D(K)$ between $\tilde{C}$ and C as the degree inequality between them in the same way as we did $I(k)$.

As shown in Fig. 15.1, the knowledge that the researcher requires—in other words, the documents that are viewed as existing concepts—is expressed by ordinary sets of keywords M_A, M_B,... on the *a priori* knowledge space. When we assume that the keyword set for one document M_S has equal weight in the description of that document, the portion R_S of all M_S that is included in the recognition concept is defined as shown in Fig. 15.7. The computer outputs the documents with an R_S value greater than that indicated by the user as requested information.

15.1.6   Results of a Simulation of the Recognition Concept Revision Process

In Fig. 15.8, it is assumed that attributes 1–50 of the *a priori* knowledge space are distributed at regular intervals on a one-dimensional axis, and the figure shows the process of the computer's recognition of recognition concept $C = \{20, 21, \ldots 33\}$. First, the researcher's initial specification $x_s = 25$ gives rise to initial recognition $\tilde{C}_0 = F_{25}(= F_{x_s})$, with ambiguity index $I(0) = 0.350$ and degree of inequality with C, $D(0) = 1.48$. Following the basis for choice in 15.1.3, the computer exhibits the first ($k = 1$) optimal query attribute to the researcher. Next, since $18 \notin C$, the researcher responds with "no." The computer's recognition is revised to $\tilde{C}_1$, and the values of both I and D are reduced. In the same way, 22 is chosen as the query attribute. Since $22 \in C$, the researcher responds with "yes," and the recognition is revised to $\tilde{C}_2$. In this simulation, we can tell from (6) in the figure that there are $K = 23$ revisions, by means of which the computer's recognition concept $\tilde{C}_{23}$ is almost the same

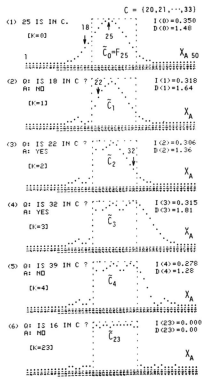

Fig. 15.8. Simulation of Recognition Concept Revision Process on a One-Dimensional *A Priori* Knowledge Space (Nakamura, K., and Iwai, S.: "Topological Fuzzy Sets as a Quantitative Description of Analogical Inference and Its Application to Question-Answering Systems for Information Retrieval," *IEEE Trans. on Systems, Man, and Cybernetics,* SMC-12, 2, March/April 1982.)

as C. In this simulation,

$$f_{x_k}(x) = \exp(-\alpha|x - x_k|^2)$$

is used as the fuzzification function for quantifying the estimation strength, and the adjustment of the estimation strength for query attribute x_i in the boundary neighborhood of C is carried out by changing the value of α. Here, the range $f_{x_k}(x) < 0.05$, $\bar{f}_{x_k}(x) > 0.95$ is defined as the neighborhood of x_k.

15.2 PROTOTYPE DOCUMENT RETRIEVAL SYSTEM

The prototype system is one that has *a priori* knowledge space X determined by the process in 15.1.2 from 812 keywords that are included in technical papers concerning mechanical engineering, with 5085 keyword and document pairs. The fuzzification function for quantification of estimation strength is the same as that used in the simulation of 15.1.6, $f_{x_k}(x) = \exp(-\alpha|x - x_k|^2)$, and the initial selected value for α, α_0, is determined by the user, as can be seen from the [PROC.1]Q:ALPHA?, A:.006 (Q is the computer's query, A the user's answer) query in Fig. 15.9. As we have already said, in actual information retrieval, the researcher cannot clearly state the concept he requires. However, in order to evaluate the characteristics of this system quantitatively, we let the researcher's required concept C be given by the union set of several documents M. Fig. 15.9 is a test example in which the required concept C is the same as the single document M_{1447} ("Machining Center Does 1500 Operations," with four keywords for descriptive attributes: MACHINING CENTER, NUMERICAL CONTROL, CUTTING, and CRANK SHAFT). In [PROC.2] the computer creates $\tilde{C}_0$ using the researcher's answer $x_s =$ "machining center" to its request for an input for an initial attribute x_s. As a result of inference, the computer outputs 12 keywords that have a degree of inclusion in $\tilde{C}_0$, which is on the *a priori* knowledge space, of 0.1 or greater [PROC.3] shows the recognition revision process in which the computer chooses a keyword using the query attribute selection basis in 15.1.3, and the researcher answers "yes" or "no" as to whether it is included in his own request concept $C(= M_{1447})$. In this system, revision ends when the ambiguity index $I(k)$ for recognition $\tilde{C}_k$, which was discussed in 15.1.3, is reduced to half of the maximum value it has had up to the present. In this example, $I(11) = 0.03177 < 1/2 \max I(k) = 1/2 \times I(0) = 1/2 \times 0.06572$, so the revision process ends after $k = 11$ times. As is shown in the figure, most of the keywords are close to either 0 or 1, and we know that the ambiguity of $\tilde{C}_{11}$ is small enough, which means it is close to being an ordinary set. We know that in the revision process, if the number of keywords in the output (degree of inclusion is 0.1 or greater) increases and the recognition range is expanded, the value of I increases, and conversely, if the number of keywords decreases and the recognition range is shrunk, the value of I decreases. Thus I is not monotonic, but is ultimately connected with 0 as we can see from what was mentioned in 15.1.5. In [PROC.4], the computer asks to what degree recognition concept $\tilde{C}_{11}$ is required in the documents (expressed as a ordinary set of keywords on the *a priori* knowledge space) to be output. When the researcher answers with an R value of 0.9, the output is three documents, M_{1416}, M_{1447}, and M_{1448}, and when the answer is 0.5, the output contains 15 documents. M_{1447} is requested concept C itself, but M_{1416} and M_{1448} include the keywords NUMERICAL CONTROL and MACHINING CENTER, so

```
[PPOC.1] O: ALPHA ?
         A: 0.006

[PPOC.2] O: WHAT KIND OF INFORMATION DO YOU NEED ?
         PLEASE INPUT A KEYWORD OF THE INFORMATION.
         A: MACHINING CENTRE

                 I*0*=0.06572
                   KEYWORD          MEMBERSHIP FUNCTION
                 RECIPROCATING COMPRE0.61
                 CONTACT RATIO        0.10
                 MACHINING            0.14
                 CRANK SHAFT          0.74
                 MACHINE TOOL         0.56
                 LUBRICATION          0.45
                 NUMERICAL CONTROL    0.88
                 CUTTING              0.21
                 CHATTERING           0.11
                 SURFACE ROUGHNESS    0.25
                 MILLING              0.68
                 MACHINING CENTRE     1.00

[PROC.3]*1*O: IS RECIPROCATING COMPRESSOR   *2*O: IS LUBRICATION              *3*O: IS NUMERICAL CONTROL
         CONTAINED IN THE INFORMATION ?         CONTAINED IN THE INFORMATION ?      CONTAINED IN THE INFORMATION ?
         A: NO                                  A: NO                               A: YES
           I*1*=0.05753                           I*2*=0.04811                        I*3*=0.04275
           CONTACT RATIO      0.10                CONTACT RATIO      0.10             CONTACT RATIO      0.10
           MACHINING          0.14                MACHINING          0.14             MACHINING          0.14
           CRANK SHAFT        0.12                CRANK SHAFT        0.12             CRANK SHAFT        0.18
           MACHINE TOOL       0.56                MACHINE TOOL       0.56             MACHINE TOOL       0.70
           LUBRICATION        0.45                NUMERICAL CONTROL  0.88             NUMERICAL CONTROL  1.00
           NUMERICAL CONTROL  0.88                CUTTING            0.21             CUTTING            0.21
           CUTTING            0.21                CHATTERING         0.11             CHATTERING         0.11
           CHATTERING         0.11                SURFACE ROUGHNESS  0.25             SURFACE ROUGHNESS  0.25
           SURFACE ROUGHNESS  0.25                MILLING            0.68             MILLING            0.68
           MILLING            0.68                MACHINING CENTRE   1.00             MACHINING CENTRE   1.00
           MACHINING CENTRE   1.00

      *4*O: IS CRANK SHAFT               *5*O: IS CONTACT RATIO           *6*O: IS MACHINING
         CONTAINED IN THE INFORMATION ?         CONTAINED IN THE INFORMATION ?      CONTAINED IN THE INFORMATION ?
         A: YES                                 A: NO                               A: NO
           I*4*=0.04086                           I*5*=0.04025                        I*6*=0.03907
           CONTACT RATIO      0.10                MACHINING          0.14             CRANK SHAFT        1.00
           MACHINING          0.14                CRANK SHAFT        1.00             MACHINE TOOL       0.70
           CRANK SHAFT        1.00                MACHINE TOOL       0.70             NUMERICAL CONTROL  1.00
           MACHINE TOOL       0.70                NUMERICAL CONTROL  1.00             CUTTING            0.21
           NUMERICAL CONTROL  1.00                CUTTING            0.21             CHATTERING         0.11
           CUTTING            0.21                CHATTERING         0.11             SURFACE ROUGHNESS  0.25
           CHATTERING         0.11                SURFACE ROUGHNESS  0.25             MILLING            0.68
           SURFACE ROUGHNESS  0.25                MILLING            0.68             MACHINING CENTRE   1.00
           MILLING            0.68                MACHINING CENTRE   1.00
           MACHINING CENTRE   1.00

      *7*O: IS MACHINE TOOL              *8*O: IS CUTTING                 *9*O: IS POSITIONING
         CONTAINED IN THE INFORMATION ?         CONTAINED IN THE INFORMATION ?      CONTAINED IN THE INFORMATION ?
         A: NO                                  A: YES                              A: NO
           I*7*=0.03292                           I*8*=0.04482                        I*9*=0.04263
           CRANK SHAFT        1.00                POSITIONING        0.17             RECIPROCATING COMPRE0.10
           NUMERICAL CONTROL  1.00                RECIPROCATING COMPRE0.10            CRANK SHAFT        1.00
           CUTTING            0.21                CRANK SHAFT        1.00             AUTOMATIC LATHE    0.64
           CHATTERING         0.11                AUTOMATIC LATHE    0.64             NUMERICAL CONTROL  1.00
           SURFACE ROUGHNESS  0.25                NUMERICAL CONTROL  1.00             NC LATHE           0.13
           MILLING            0.68                NC LATHE           0.13             CUTTING            1.00
           MACHINING CENTRE   1.00                CUTTING            1.00             THERMAL DEFORMATION0.17
                                                  THERMAL DEFORMATION0.17            CHATTERING         0.11
                                                  CHATTERING         0.11            SURFACE ROUGHNESS  0.25
                                                  SURFACE ROUGHNESS  0.25            MILLING            0.68
                                                  MILLING            0.68            MACHINING CENTRE   1.00
                                                  MACHINING CENTRE   1.00

      *10*O: IS SURFACE ROUGHNESS        *11*O: IS MILLING
         CONTAINED IN THE INFORMATION ?         CONTAINED IN THE INFORMATION ?
         A: NO                                  A: NO
           I*10*=0.04332                          I*11*=0.03177
           RECIPROCATING COMPRE0.10               RECIPROCATING COMPRE0.10
           CRANK SHAFT        1.00                CRANK SHAFT        1.00
           AUTOMATIC LATHE    0.64                AUTOMATIC LATHE    0.64
           NUMERICAL CONTROL  1.00                NUMERICAL CONTROL  1.00
           NC LATHE           0.13                NC LATHE           0.13
           CUTTING            1.00                CUTTING            1.00
           THERMAL DEFORMATION0.17                THERMAL DEFORMATION0.17
           CHATTERING         0.11                CHATTERING         0.11
           MILLING            0.42                MACHINING CENTRE   1.00
           MACHINING CENTRE   1.00

[PPOC.4]*1*O: HOW MUCH AMOUNT OF THE INFORMATION    *2*O: HOW MUCH AMOUNT OF THE INFORMATION
         SHOULD BE CONTAINED IN THE DOCUMENT ?            SHOULD BE CONTAINED IN THE DOCUMENT ?
         PLEASE INPUT THE AMOUNT A: PROPORTION.           PLEASE INPUT THE AMOUNT A: PROPORTION.
         A: 0.9                                           A: 0.5

[OUTPUT]     NUMBERS OF THE DOCUMENTS YOU NEED           NUMBERS OF THE DOCUMENTS YOU NEED
             1416                                        1302    1396
             1447                                        1339    1403
             1448                                        1341    1406
               -- END OF DATA                            1342    1416
                                                         1347    1417
                                                         1347    1447
                                                         1353    1448
                                                         1392      -- END OF DATA
```

Fig. 15.9. Dialog-type Retrieval in the Prototype Document Retrieval System (Nakamura, K., and Iwai, S.: "Topological Fuzzy Sets as a Quantitative Description of Analogical Inference and Its Application to Question-Answering Systems for Information Retrieval," *IEEE Trans. on Systems, Man, and Cybernetics*, SMC-12, 2, March/April 1982.)

they are the same as C in part. However, the contents of M_{1416} ("Two Typical Applications of CN") and M_{1448} ("The Economical Efficiency of Keyboard Programming") are the same as M_{1447} in that they deal with numerical control in human-machine centers. Therefore they are suitable to the requirements of the researcher. With the increase in the value of R, documents more distant from the request concept are included in the output.

15.3 CHARACTERISTICS OF REQUEST CONCEPTS AND RECOGNITION EFFICIENCY

Recognition of request concepts in this system is achieved by means of estimations corresponding to the distance between keywords from input information on the keyword space in the computer's *a priori* knowledge. Therefore, its efficiency is influenced by the extent and coherency of request concept C, which is its object. Evaluation of the recognition efficiency can be done using the number of dialogs k_e before termination of recognition revision (k when $I(k)$ has been reduced to half of its maximum value; the smaller, the faster the recognition) and the degree of inaccuracy $D(k_e)$ of the recognition concept and the request concept at the time of termination (the Euclidean distance between $\tilde{C}_{ke}$ and C; the smaller, the more accurate the recognition).

15.3.1 Extent of Request Concept and Estimation Strength

Let us consider three request concepts that differ in extent.

C_I Document M_{406} (keywords: BOUNDARY LAYER, FORCED CONVENTION, FREE CONVECTION, FLAT PLATE), $C_I^\# = 4$.

C_{II} $M_{390} \cup M_{408} \cup M_{409} \cup M_{410} \cup M_{469}$ (all including the keyword BOUNDARY LAYER), $C_{II}^\# = 8$.

C_{III} Union set of keywords for 12 documents concerning "combustion systems," $C_{III}^\# = 23$.

Figure 15.10 shows the relationships between the values of α for fuzzification function $f(x) = \exp(-\alpha|x - x_k|^2)$ and the recognition efficiency values $(k_e, D(k_e))$. In the cases of C_I, C_{II}, and C_{III}, an optimal value for α, α_0 (the closest value to the point of origin in the k_e, $D(k_e)$ plane) exists, and that value is larger (distance from the point of origin is less) when the extent $(C^\#)$ of the request concept is small, so we know that the efficiency of recognition itself is large

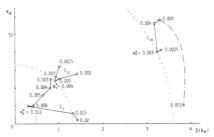

Fig. 15.10. Extent of Request Concept and Recognition Efficiency (Nakamura, K., and Iwai, S.: "Topological Fuzzy Sets as a Quantitative Description of Analogical Inference and Its Application to Question-Answering Systems for Information Retrieval," *IEEE Trans. on Systems, Man, and Cybernetics*, SMC-12, 2, March/April 1982.)

(distance from the point of origin is small). In the case of C_{III}, in which the extent of the request concept is large, k_e jumps when α_0 increases from 0.005 to 0.008 (when the width of the fuzzification function is reduced). This means that the range of the estimation is too narrow when compared to the extent of C_{III} and that recognition concept revision was stopped at a stage where one part of C_{III} (concretely, a part that covered 8 of the 23 keywords for C_{III}) was recognized.

15.3.2 Request Concept Coherency and Estimation Strength

Since the fuzzification function that determines the estimation strength is bell-shaped, we can predict that the recognition efficiency will improve in cases in which the concept requested by the researcher provides a group with a small dispersion in the keyword space. Let us now consider the maximum range, as shown in Fig. 15.11 from one keyword x_j in keyword set C, which expresses the request concept. Of the keywords included in C, the one with the minimum maximum range, x_c, is called the *center* of C. When we let the set of keywords

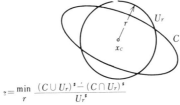

$$z = \min_r \frac{(C \cup U_r)^s - (C \cap U_r)^s}{U_r^s}$$

Fig. 15.11. Center x_c of Request Concept and Concept Coherency Measure z

within the range γ of x_c be U_γ, the coherency of the request concept can be expressed by

$$z = \min_\gamma \frac{(C \cup U_\gamma)^{\#} - (C \cap U_\gamma)^{\#}}{U_\gamma^{\#}}$$

($\#$ being the number of elements in the set).

Here the extent—that is the number of keywords included—is just about the same, but let us consider the following three cases in which the values for z differ:

C_a C_{II}, $C_a^{\#} = 8$, $z = 0.200$;

C_b keyword set of eight documents that have the keyword MACHINE TOOL in common, $C_b^{\#} = 9$, $z = 0.782$;

C_c set of eight randomly chosen keywords, $C_c^{\#} = 8$, $z = 0.989$.

Figure 15.12 shows the efficiency for C_a, C_b, and C_c using the optimal α_0 value for each, as in 15.3.1. We can see that, as was predicted, the smaller the value of z (more coherent concept), the closer it is to the point of origin (greater recognition efficiency).

As with documents M_A, $M_B, \ldots$, one recognition, $\tilde{C}_{ke}$, can be reflected in the distance between keywords in the *a priori* knowledge space, which was described as existing concepts in 15.1.1. The results are that the distance between keywords included in $C (\cong \tilde{C}_{ke})$ shrinks and that between keywords not included in C opens up. In other words, this kind of adjustment of the *a priori* knowledge space improves the coherency of requested concept C when recognition of C piles up, and we predict that it will have the effect of improving recognition efficiency. Since Fig. 15.13 shows the relationship between the number of recognitions, coherency of $C(z)$ and the value for evaluating recognition efficiency $(k_e, D(k_e))$, it serves as proof of what is stated above.

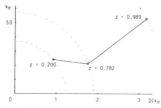

Fig. 15.12.   Request Concept Coherency and Recognition Efficiency (Nakamura, K., and Iwai, S.: "Topological Fuzzy Sets as a Quantitative Description of Analogical Inference and Its Application to Question-Answering Systems for Information Retrieval," *IEEE Trans. on Systems, Man, and Cybernetics*, SMC-12, 2, March/April 1982.)

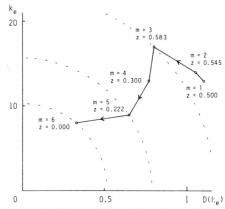

Fig. 15.13. Adjustment of *A Priori* Knowledge Space Using Recognition Experience and Its Effects on Recognition Efficiency (Nakamura, K., and Iwai, S.: "Topological Fuzzy Sets as a Quantitative Description of Analogical Inference and Its Application to Question-Answering Systems for Information Retrieval," *IEEE Trans. on Systems, Man, and Cybernetics*, SMC-12, 2, March/April 1982.)

15.4   THE ROLE OF *A PRIORI* KNOWLEDGE AND INTELLIGENT INTERFACES

For solving the complex problems in the coming era of diversification, more and more is going to be required of man's creative abilities. The range of knowledge that each person can understand is limited. Dialogue with a vast amount of knowledge, necessary for solving these problems, is made possible first by the integration of the knowledge of many fields. It can be said that in order for this dialogue to be carried out smoothly and with high efficiency, support for intelligent computer operations, in which the dialogue level is raised by the introduction of new computer technology (intelligent interfaces) at the point of contact of humans and the information in the computer's memory, is indispensable.

The intelligent interface introduced here is one that offers complete information for the user's incomplete, ambiguous requests. It does this by taking the researcher's fragmentary information, that is, information in the form of points, and producing information in the form of planes using these points as centers of a spread. All of this is based on the computer's *a priori* knowledge, which is organized to use effectively a vast amount of accumulated knowledge in the form of keyword spaces. Recently, various types of consultation systems have been proposed under the name of knowledge information processing systems. For example, a system that serves in the place of a travel agency makes up a complete travel plan that can actually be used just by receiving an outline of what the user (traveler) wants.[9] What is in the database for this system is not just the vast amount of data in documents about all the cities of the world and information about transportation service networks; it also holds the consultant's indispensable knowledge of how to use this data to meet the traveler's requirements and how the design of a typical travel plan should progress, including things like visa procedures. By means of the knowledge for controlling this information, a smooth dialogue is possible. The knowledge that this consultant must have corresponds to the *a priori* knowledge for putting this accumulated knowledge to use that we have discussed in this chapter. Thus the establishment of expressions for *a priori* knowledge that are appropriate for the problem range can be considered to be the most important problem in research on intelligent interfaces.

REFERENCES

(1) Nakamura, K., and Iwai, S., "Topological Fuzzy Sets as a Quantitative Description of Analogical Inference and Its Application to Question-Answering Systems for Information Retrieval," *IEEE Transactions on Systems, Man and Cybernetics*, SMC-12, pp. 193–204 (1982).

(2) Quillian, M. R., "Semantic Memory," in Minsky, M., ed., *Semantic Information Processing*, pp. 227–270, M. I. T. Press, Cambridge, Mass. (1968).

(3) Minsky, M., "A Framework for Representing Knowledge," in Winston, P., ed., *The Psychology of Computer Vision*, pp. 211–277, McGraw-Hill (1975).

(4) Atkinson, R. C., and Estes, W. K., "Stimulus Sampling Theory," in Luce, R. D., Bush, R. R., and Galanter, E., eds., *Handbook of Mathematical Psychology*, **2**, pp. 121–268, John Wiley, New York (1963).

(5) Tversky, A., "Features of Similarity," *Psychological Review*, **84**, 4, pp. 327–352 (1977).

(6) Sjöberg, L., "A Cognitive Theory of Similarity," Göteborg Psychological Reports, **10** (1972).

(7) Doyle, L. B., "Indexing and Abstracting by Association," *Amer. Documentation*, **13**, pp. 378–390 (1962).

(8) Kaufmann, A., *An Introduction to the Theory of Fuzzy Subsets*, **I**, Academic Press, Cambridge, Mass. (1975).

(9) Bobrow, D. G., Kaplan R. M., Kay, M., Norman, D. A., Thompson, H., and Winograd, T., "GUS, A Frame-Driven Dialog System," *Artificial Intelligence*, **8**, pp. 155–173 (1977).

Chapter 16

EXPERT SYSTEM FOR DAMAGE ASSESSMENT

16.1 EXPERT SYSTEMS AND FUZZY KNOWLEDGE

As discussed in Section 9.4, a great deal of attention is being paid to expert systems as a method for the systemized use of knowledge in a specific problem domain. The academic research area for the basic technology for the construction of these expert systems is called *knowledge engineering*, and what forms the basis of knowledge engineering is research into artificial intelligence. In other words, knowledge engineering can be called practical artificial intelligence, since it is an area oriented toward applications of artificial intelligence, and expert systems are concrete products of the loading-up of knowledge of specialists. Since the mid-1970s advent of the famous expert system, MYCIN (a system for supporting diagnosis and treatment for bacteria in the blood developed at Stanford University), a large number of expert systems have been constructed. Table 16.1 shows most of the areas in which applications of expert systems have been made.

The basic structure of an expert system is shown in Fig. 16.1. The main characteristic is its separation into a knowledge base to store modular knowledge and an inference mechanism for using it.

Table 16.1.   Major Applications of Expert Systems

Type of Work	Examples
Analysis, Interpretation	signal analysis, elucidation of chemical structures, understanding voices and images
Diagnosis	medical diagnosis, circuit fault diagnosis, mechanical fault diagnosis, plant diagnosis, financial diagnosis, program diagnosis, (including action and repair indications)
Supervision, Management	plant operation, hospital supervision, project management, financial management
Prediction	grain harvest estimation, stock price and exchange rate prediction, weather forecasts, military model forecasts (including simulation technology)
Planning	experiment planning, travel planning, schedule coordination, project planning
Instructions, Education	CAI
Design	VLSI circuit design, architectural design, materials design, mechanical design, program design, drawing up budget proposals

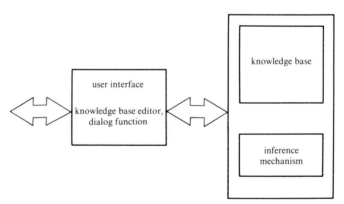

Fig. 16.1.   Expert System Structure

The following items are recognized subjects in knowledge engineering and expert systems.

Knowledge Representation Methods
How should knowledge be represented?

Knowledge Utilization Methods
How should knowledge be coordinated and used in order to attain the goal?

Knowledge Acquisition and Management
When the system grows, how can knowledge be automatically or semiautomatically integrated into it? In addition, how can knowledge base be managed so as to remove contradictions and preserve consistency?

User Interface
How can we make an efficient, human-like user interface for use, acquisition, and management? (This includes graphic and natural language processing.)

Besides these subjects, of course, accumulation and arrangement of specialist knowledge in the individual areas are also important topics.

The method of expressing knowledge has an intimate connection with the inference mechanism and is the nucleus for the construction of an expert system. Some representative knowledge representation methods developed up to now in the field of knowledge engineering are as follows:

(1) production system (rule expression)
(2) blackboard model
(3) semantic network
(4) frame
(5) predicate logic
(6) object.

One object area in which expert systems are expected to be effective is that of ill-defined problems (problems in which it is either difficult or impossible strictly to define a method of approach). In most cases, either the clear knowledge of ill-defined problems cannot be obtained, or completion of the knowledge set must be approached gradually. In other words, there are many cases in which knowledge is ambiguous.

On an everyday basis, people make judgments and solve problems in an environment with ambiguous knowledge. The basis of knowledge engineering is symbolic inference by means of symbol manipulation; however, since it is close to human thinking, it goes beyond the framework of simple symbolic inference. Accordingly, we have to find expressions for "ambiguous knowledge" and an inference mechanism that will work in that environment.

When we say "ambiguous," the meaning of the word itself is ambiguous and vague. Just as knowledge must be formulated and written down for a computer to understand it, ambiguity must take some form before it can be dealt with technically. Therefore, ambiguity that is dealt with in knowledge engineering is classified and written down as follows.[1]

(1) nondeterminism
(2) multiple meanings
(3) uncertainty
(4) incompleteness
(5) fuzziness or imprecision

(3) uncertainty and (5) fuzziness have a particularly close relationship with the main subject of this book, and systems that handle knowledge with fuzziness have been created even in the field of knowledge engineering.[2,3] Knowledge engineering methods for dealing with uncertainty are noted in reference 4).

In this chapter we will introduce an expert system for building damage assessment called SPERIL. In 1979, J. T. P. Yao (civil engineering) and K. S. Fu (electrical engineering) of Purdue University in Indiana began a research project using National Science Foundation funds for the development of a damage assessment system for buildings that had been subjected to earthquake vibrations as the main object. Through this project, both Ishizuka's SPERIL-I[5−7,12] and Ogawa's SPERIL-II[8−10] were developed.

The current state of structural damage assessment engineering and its role have been described by Yao.[11,12] For example, it is easy to identify a small number of buildings such as those that have suffered partial collapse after an earthquake, but it is not easy to grasp accurately the degree of damage to a large number of other buildings or to judge accurately the necessity and methods for repair. At present this is being carried out by a few specialists with experience in damage assessment, and the passing-on of methods for judging mainly takes the form of apprentice relationships. Several methods for structural damage assessment have been proposed up to now, and in related research, several methods for evaluating durability in disasters have been developed. However, a reasonable, systematic method for damage assessment has yet to be established.

At first Fu and Yao suggested handling this problem by means of pattern recognition theory. In pattern recognition, it is necessary to express the patterns for the problem by means of a mathematical model, using either a statistical method or a grammatical/structural method, and for this it is necessary to have statistical knowledge of the patterns taken as objects. In complex problems in which it is difficult to express things clearly, as in the damage assessment here or in medical diagnosis[13,14], this kind of knowledge often cannot be obtained. In addition, since there is great variety in the conditions of the buildings that are the objects, there are many difficulties in using pattern recognition methods based on a standard model. Therefore, since 1980, damage assessment systems have been constructed using the framework of expert systems in which knowledge of specialists is gathered and the answers are derived through inference. A reasonable inference system for cases in which the knowledge carries uncertainty or fuzziness has been developed as the basic theory of system construction.[15–17] This is an extension of Dempster–Shafer probability theory[18–20] to fuzzy sets. We will give an outline of the SPERIL expert system for building damage assessment and this inference method here.

16.2 EXPRESSION OF PROBLEMS THAT CONTAIN UNCERTAINTY

Concerning problems that include a complex variety of conditions, it is more efficient to gather information in a fragmentary way. This idea is emphasized in knowledge engineering, and it can also be seen from the viewpoint of the problem reduction method. In problem reduction, the problem is divided into a number of simple parts, and these partial problems are further divided into simpler problems. Therefore, the problem is expressed hierarchically. Just as the overall problem has a final goal that must be achieved, each of the smaller problems has a subgoal that must be attained using obtainable information.

If a production system is used, fragmentary knowledge that is used to infer the upper-level subgoals or the final goal from the lower-level subgoals and observational evidence is expressed using the following form of production rules.

Rule   IF: X

THEN A

Here the IF and THEN parts are called the proposition (or condition) and action (or conclusion) respectively. The basic function of the rules is to update the state of the indicated subgoal or final goal in the action part, if the proposition is satisfied. In judgment problems in the real world, there are cases

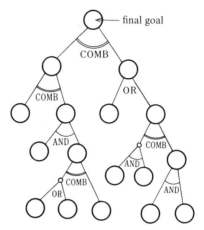

Fig. 16.2.   AND/OR/COMB Tree for Problems Accompanying Uncertainty

when circumstances are not clear enough to express values of true or false. There are two types of uncertainty here. The first is the uncertainty that accompanies observed data or evidence, and the second is that which accompanies expressed knowledge. In the development of SPERIL, it was necessary to use data and knowledge with this kind of uncertainty effectively.

In judgment problems that are accompanied by uncertainty, the combination relation expressed by COMB is needed in addition to the AND/OR relations between subproblems. The combination relation indicates a relation between subproblems such as when a goal is supported independently by two or more pieces of evidence. In this way, the problem is expressed by a AND/OR/COMB tree (generally a graph) like that in Fig. 16.2.

Inference for AND/OR relations is easy; min or max operations are employed with the certainty measures* for each, and there is no other appropriate method. Therefore, it is necessary to determine a reasonable inference method for the COMB relation along with the certainty measure. A reasonable inference method means that at least a correct answer will be derived as long as knowledge is correct.

Let us consider the basic circumstances as shown in Fig. 16.3, when two independent evidential states X and Y for a given goal or hypothesis A have been observed or obtained from preceding inferences. Here we will assume

---

* For the mathematical expression of certainty, there is the probability quantity that follows Bayes's rules, the Dempster–Shafer probability quantity, and truth value from fuzzy theory, but we use *probability measure* here as a generic name.

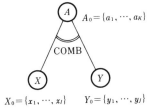

$$A_0 = \{a_1, \cdots, a_K\}$$

$$X_0 = \{x_1, \cdots, x_I\} \qquad Y_0 = \{y_1, \cdots, y_J\}$$

Fig. 16.3.   Inference by Means of Combining Different Pieces of Evidence

that the knowledge base has the following rules:

Rule 1   IF: X_1

THEN A_{11} with C_1

Rule 2   IF: Y_1

THEN A_{21} with C_2

Here we let X_1, Y_1, A_{11} and A_{21} express subsets of finite complete sets X_0, Y_0 and A_0, respectively. The problem then is how to infer the certainty of the goal or hypothesis states.

In MYCIN, an intuitive combination function was used to attain this goal.[21] The MYCIN method had a large influence on the handling of uncertain knowledge in expert systems, but it does not have a rational background. Duda, Hart, Nilsson, *et al.* have modified Bayes's rules for use in expert system reasoning and proposed a subjective Baysian method.[22] In contrast, the importance of Dempster–Shafer theory in the inference for expert systems that is accompanied by uncertainty has been recognized by Ishizuka and others.[23–25] Dempster–Shafer theory allows rational handling of uncertainty that is connected with subjectivity, and the inference mechanism for SPERIL was designed based on an extension of this theory.

If an inference process for COMB relations as well as AND/OR relations can be defined, the certainty measure can be propagated through the hierarchical inference network. In the end, the certainty of the final goal can be determined and a rational answer obtained for the judgment.

16.3   DEMPSTER–SHAFER THEORY AND ITS EXTENSION TO FUZZY SETS

The main criticism against handling subjective uncertainty by means of probability quantities that follow Bayes's rules (for convenience we will call this Baysian probability here) is that Baysian probability cannot handle ignorance effectively. In other words, if we let $P(A)$ be the probability of A, the relation

$$P(A) + P(\bar{A}) = 1 \tag{16.1}$$

($\bar{A}$ being the complement of A) is required in Bayes's theory, so there is no distinction between lack of belief and disbelief.

In 1967 Dempster proposed new ideas, giving them the names *upper* and *lower probabilities*, for handling the uncertainty connected with subjectivity, for which Baysian probability was inappropriate.[18] Shafer refined Dempster's theory, and, in order to endow them with a subjective meaning, changed the names of the original upper and lower probabilities to *belief function* and *plausibility*, respectively.[19] These are defined through the basic probability.[19] One important point is that Dempster and Shafer's idea on probability includes Baysian probability as a special case, as we will explain.

First, the Dempster–Shafer theory does not adhere to the additivity rule of equation (16.1), which seems natural in Baysian probability. Shafer's basic probability $m(a_i)$ is kept within subset A_i, but its image can be depicted as a semimobile probability mass that can move freely among the points within A_i, as shown in Fig. 16.4. Since this image is the key to understanding, we will add an explanation using Fig. 16.4 as an example. In Fig. 16.4, we will let $a_1 - a_6$ be elements that might possibly be obtained, and their complete set be $A_0 = \{a_1, \ldots, a_6\}$. If we consider the basic probability $m(\{a_4, a_5\})$, which is apportioned to subset $\{a_4, a_5\}$ in Fig. 16.4, as mass of 0.3, this mass can move to either a_4 or a_5 and stay there. In addition, it can break up as 0.1 and 0.2 and remain in both elements. In the same way the basic probability $m(\{a_1, \ldots, a_6\})$ apportioned to the complete set can move to any element. In fact, the basic probability apportioned to the complete set expresses the degree of ignorance. Basic probabilities $m(\{a_1\})$ and $m(\{a_4\})$ which remain in elements a_2 and a_4, respectively, cannot move and are static probability masses.

If we let A_0 be a finite complete set and $A_i (i = 1, 2, \ldots)$ a subset of it, the

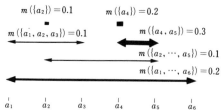

Fig. 16.4.   Image $A_0 = |a_1, \ldots, a_s|$ of Dempster–Shafer's Basic Probability $m(A_i)$

basic probability $m(A_i)$ $(i = 1, 2, \ldots)$ takes a $[0, 1]$ value and satisfies the following condition:

$$\begin{cases} m(\varnothing) = 0 & (\varnothing: \text{empty set}) \\ \sum_{A_i \subseteq A_0} m(A_i) = 1 \end{cases} \tag{16.2}$$

When $m(A_i) > 0$, A_i is called a focal element.

Using the basic probability, the lower probability is defined as

$$P_*(A_i) = \sum_{A_j \subseteq A_i} m(A_j). \tag{16.3}$$

In other words, it is the sum of the basic probabilities held within subset A_i. On the other hand, the upper probability is defined as

$$\begin{aligned} P^*(A_i) &= 1 - P_*(\bar{A}_i) \\ &= 1 - \sum_{A_j \subseteq \bar{A}_i} m(A_j). \end{aligned} \tag{16.4}$$

In other words, it is the total sum of all basic probabilities that could possibly enter A_i even a little. For Fig. 16.4, the upper and lower probabilities for subset $\{a_4, a_5\}$ come out as follows:

$$P_*(\{a_4, a_5\}) = m(\{a_4\}) + m(\{a_4, a_5\}) = 0.5 \tag{16.5}$$

$$P^*(\{a_4, a_5\}) = m(\{a_4\}) + m(\{a_4, a_5\}) + m(\{a_2, \ldots, a_5\})$$
$$+ m(\{a_1, \ldots, a_6\}) = 0.8. \tag{16.6}$$

In other words, the lowest probability for $\{a_4, a_5\}$ is 0.5 in this circumstance, and depending on the case could go as high as 0.8. We cannot say where it is between 0.5 and 0.8.

The fact that the basic, lower, and upper probabilities do not satisfy additivity can be easily understood. If the basic probability were always static as with $m(\{a_2\})$ and $m(\{a_4\})$, the Dempster–Shafer idea would degenerate to Baysian probability. This is how Baysian probability is included as a special case.

Dempster's Combination Rule

One important rule is Dempster's rule of combination. This gives the method for combining the basic probabilities that are inferred from independent evidence. Let m_1 and m_2 be basic probabilities obtained from independent evidence, and A_{1i} and A_{2j} $(i, j = 0, 1, 2, \ldots)$ be the focal points for each. Dempster's rule of combination can make the following consolidation, and we come up with a new basic probability:

$$m(A_k) = \frac{\displaystyle\sum_{A_{1i} \cap A_{2j} = A_k} m_1(A_{1i}) m_2(A_{2j})}{1 - \displaystyle\sum_{A_{1i} \cap A_{2j} = \phi} m_1(A_{1i}) m_2(A_{2j})}. \tag{16.7}$$

where $A_k \neq \phi$.

As is shown in Fig. 16.5, the numerator means that the product of the basic probabilities is apportioned in the product set A_k of A_{1j} and A_{2j}. Originally this was enough, but there is a problem when the product set is the empty set. Therefore, the case of the empty set is excluded, and the whole is normalized by the denominator. When two or more basic probabilities that have been inferred from independent evidence are combined, Equation (16.7) is used successively to obtain the result.

Using this rule, application of the Dempster–Shafer theory in expert systems is easy. Let us consider the problem in Fig. 16.3 in which certainty measure for hypothetical state A is determined from Rules 1 and 2, which connect evidence conditions X and Y. One inference process is as follows. First, the lower probability $P_*(X_1)$ is calculated for the propositions in Rule 1, and is then multiplied by certainty measure C_1. This amount is allotted as the basic probability for set A_{11}, which is shown in the conclusion:

$$m(A_{11}) = P_*(X_1) \cdot C_1. \tag{16.8}$$

In the same way $m(A_{21})$ is inferred from Rule 2 and evidence Y. These allotments of probability quantity, which are connected with H, are combined using Equation (16.7).

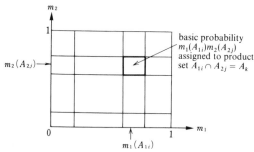

Fig. 16.5. Dempster's Combination Rule for Basic Probability

Fuzzy sets are more appropriate than crisp or nonfuzzy sets for the expression of knowledge that includes uncertainty. For example, the expressions "slight," "moderate," and "severe," which are used in SPERIL, do not make clear enough stipulations, but they are expressions that are meaningful for specialists. In order to use this kind of knowledge, Dempster–Shafer theory is extended to fuzzy sets without losing its essence. In this case, sets X_1, A_{11}, Y_1, and A_{21}, which show the restrictions of Rules 1 and 2, can be characterized by membership functions, so that they can be fuzzy sets.

In order to extend Dempster–Shafer theory, we define the degree to which fuzzy subset A_1 within complete set A_0 is included in another fuzzy subset A_2 as follows:

$$I(A_1 \subseteq A_2) = \frac{\min\limits_{a}\{1, 1 - \mu_{A_1}(a) + \mu_{A_2}(a)\}}{\max\limits_{a}\{\mu_{A_1}(a)\}}. \tag{16.9}$$

Here, $\mu_{A_1}(a)$ and $\mu_{A_2}(a)$ are the membership functions for A_1 and A_2, respectively. We define the degree to which fuzzy sets A_1 and A_2 have an intersection as follows:

$$J(A_1, A_2) = \frac{\max\limits_{a}\{\mu_{A_1 \cap A_2}(a)\}}{\min\left\{\max\limits_{a}\{\mu_{A_1}(a)\}, \max\limits_{a}\{\mu_{A_2}(a)\}\right\}}. \tag{16.10}$$

Following fuzzy theory, intersection $A_1 \cap A_2$ is determined by

$$\mu_{A_1 \cap A_2}(a) = \min\{\mu_{A_1}(a), \mu_{A_2}(a)\}. \tag{16.11}$$

The degree to which the intersection of fuzzy sets A_1 and A_2 is $\varnothing$ (the empty set) is defined by $1 - J(A_1, A_2)$.

Using the above definitions, Equations (16.3) and (16.7) can be generalized as follows:

$$P_*(A_i) = \sum_{A_j} I(A_j \subseteq A_i) \cdot m(A_j) \tag{16.12}$$

$$m(A_k) = \frac{\sum\limits_{A_{1i} \cap A_{2j} = A_k} J(A_{1i}, A_{2j}) m_1(A_{1i}) m_2(A_{2j})}{\sum\limits_{A_{1i}, A_{2j}} \{1 - J(A_{1i}, A_{2j})\} m_1(A_{1i}) m_2(A_{2j})} \quad (A_k \not= \phi). \tag{16.13}$$

Therefore, a rational inference method is defined for dealing with knowledge accompanied by uncertainty and fuzziness. (Another study on the extention of Dempster–Shafer theory to fuzzy set can be seen in[27].)

16.4 SPERIL SYSTEM

SPERIL is a rule-based expert system for damage assessment of buildings that have suffered earthquake excitation. We will outline SPERIL-I here.[5–7]

In order to consolidate various observed evidence, inference by means of Dempster–Shafer theory extended to fuzzy sets as described above is employed in SPERIL. Most of the specialist knowledge that is accumulated in the knowledge base was obtained from a top researcher on structural damage models, J. T. P. Yao of Purdue University, and part of it was obtained from the data and ideas of M. A. Sozen of the Civil Engineering Department of the University of Illinois, who is a top researcher in structural earthquake response simulation.

The sources of information useful for damage judgments are:

(1) observation of various parts of buildings, and
(2) analysis of seismographical records, which are obtained during earthquake from installations in the building.

Interpretation of these data is greatly influenced by the individual conditions of building such as construction materials, height, and design. Fragmentary knowledge is accumulated based on the framework shown in Fig. 16.6, and the knowledge base is written in the form of prescribed rules. SPERIL-I was developed using C language.

As is shown in the example in Fig. 16.7, the rule is designed in a way that is easy for both people and computers to interpret. Of the four-digit number in the first line of each rule, the first two are the rule number, and this corresponds to a node number in Fig. 16.6. In order to express knowledge with fuzzy grades, the following fuzzy subset expressions are permitted:

no
slig (slight)
mode (moderate)
seve (severe)
dest (destructive)
uk (unknown—complete set)

Membership functions for these are given, as in Fig. 16.8. The functions of the rules emphasize the basic production-rule functions; in other words, if the proposition is satisfied, the action is carried out. The action in this case is the updating of state memories called STM, which correspond to the subgoals.

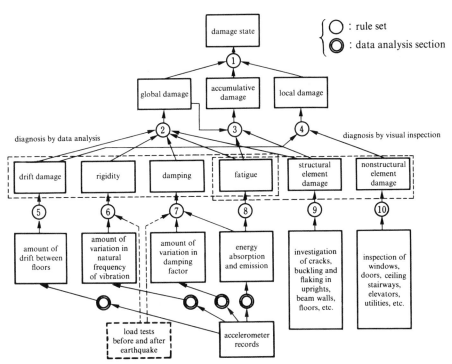

Fig. 16.6. SPERIL-I Inference Network

STM corresponds to the working memory in a production system, where input data and inferred data are saved. In SPERIL-I, STM is classified into the following four types according to content:

Type 1 Uncertainty measure of fuzzy grade

Type 2 Linguistic data

Type 3 Numerical data

Type 4 Yes-No data

When STM is accessed, this classification is referred to, and it goes on to appropriate interpretation of the rule sentences.

Since the depth of the inference network is not particularly deep, efficient inference is not given special consideration. The order for the use of rule sets is tentatively set beforehand as follows:

"05," "06," "07," "08," "09," "10," "02," "03," "04," "01".

This corresponds to bottom-up inference or data-driven inference.

```
Rule0201
       IF:MAT is    r/c
THEN IF:STI is    dest
     THEN:GLO dest 0.6
ELSE IF:STI is    seve
     THEN:GLO seve 0.6
ELSE IF:STI is    mode
     THEN:GLO mode 0.6
ELSE IF:STI is    slig
     THEN:GLO slig 0.6
ESLE IF:STI is    no
     THEN:GLO no   0.6
     ELSE:GLO uk


Rule0501
       IF:MAT is   r/c
THEN IF:ISD <=  -8.9
     THEN:DRI uk    I
ELSE IF:ISD <=   0.4
     THEN:DRI no   0.9
ELSE IF:ISD <=   0.8
     THEN:DRI slig 0.9
ELSE IF:ISD <=   1.3
     THEN:DRI mode 0.9
ELSE IF:ISD <=   2.0
     THEN:DRI seve 0.9
ELSE IF:ISD >    2.0
     THEN:DRI dest 0.9
     ELSE:DRI uk


Rule0901
       IF:MAT is    steel
THEN IF:S01 is    yes    (partial collaps)
     THEN:VST dest 1
ELSE IF:S02 Is    yes    (buckling of column)
     THEN:VST dest 0.5
      and:VST seve 0.5
ESLE IF:S03 is    yes    (buckling of girder/beam)
      or:S04 is    yes    (buckling of diagnal bracing)
      or:S05 is    yes    (deformation or loosing of joint)
     THEN:VST seve 0.9
ELSE IF:S06 is    yes    (spalling/crack on shear wall)
     THEN:VST mode 0.8
ELSE IF:S07 is    yes    (spalling/crack on exteria/interia wall)
      or:S03 is    yes    (spalling/crack on floor)
     THEN:VST mode 0.5
      and:VST slig 0.5
ELSE IF:S01 is    no
     and:S02 is    no
     and:S03 is    no
     and:S04 is    no
     and:S05 is    no
     and:S06 is    no
     and:S07 is    no
     and:S08 is    no
     THEN:VST no   I
     ELSE:VST uk
```

Abbreviations
```
   dest   destructive
   seve   severe
   mode   moderate
   slig   slight
   no     no
   uk     unknown

   r/c    reinforced concrete

   GLO    damage of global nature
   DRI    damage due to drifting
   STI    damage of stiffness
   VST    visual damage of structural member
   MAT    material of structure
   ISD    interstory drift
   S01    check items of visual structural damage for steel
    ￸
   S08
```

Fig. 16.7.   Example of Rules for SPERIL-I

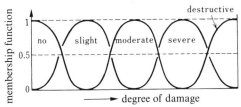

Fig. 16.8. Membership Functions for Fuzzy Subsets Used in SPERIL-I

The inference process finds rules that are associated with the current state and evaluates them. During the evaluation of the rule proposition, questions for obtaining data are generated, if the STM has not been written or there are as yet no answers. Questions are generated with reference to a question file. In order to avoid annoying unnecessary questions, a skip path is established in the control flow for cases in which there is no possibility of starting an action phrase. Accordingly, the minimum number of questions necessary for inference are generated.

After one rule has been processed, the result is used in updating the specified STM. The updating of Type I STM is carried out by means of extended Dempster–Shafer theory, assuming the inference is from independent pieces of evidence. Final judgments are based on lower probabilities of the fuzzy subsets of the final goal, which correspond to the damage condition. In cases in which a fixed threshold (0.2) is not obtained for any fuzzy subset, "no appropriate answer" is chosen.

Therefore, the answer is one of the following:

(1) no damage
(2) slight damage
(3) moderate damage
(4) severe damage
(5) destructive damage
(6) no appropriate answer

SPERIL-I has not as yet included knowledge about repairs and reinforcement.

After SPERIL-I, development of SPERIL-II was implemented.[8–10] In SPERIL-II, an extended frame knowledge representation, which had functions for handling Dempster–Shafer theory and fuzzy sets, were developed using Lisp and an improved development environment. At the same time, there was greater substance in related knowledge because of cooperative research by a private structural consulting company, and the knowledge base was reconstructed.

As far as functions are concerned, performance evaluations like the recognition rate in pattern recognition cannot be obtained because of the small number of examples. However, evaluations are obtained from the judgment of specialist J. T. P. Yao; there is nothing at present that contradicts his judgment. At the same time, there are suggestions that effort must be made to make the knowledge base more complete.

Based on the development of SPERIL, Ishizuka et al. constructed a practical consulting system for assessing earthquake damage for non-life insurance companies in Japan.[26] However, since important judgments as to whether full or half insurance payments are to be made must be made in these consultations, uncertain knowledge and fuzzy knowledge representations are not appropriate and thus have been eliminated. Nucleus knowledge is brought from a handbook for damage assessment. The knowledge base is constructed from this knowledge and knowledge of experienced people. It also provides a user interface that employs images and graphics that show the damage and offer reference examples.

REFERENCES

(1) Ishizuka, M., "Representation and Utilization of Ambiguous Knowledge," *Joho Shori* (Journal of IPS of Japan), **26**, pp. 1481–1486 (1985) (in Japanese).

(2) Whalen, T., and Schott, B., "Issues in Fuzzy Production Systems," *International Journal of Man-Machine Studies*, **19**, pp. 57–71 (1983).

(3) Prade, H., "A Conceptual Approach to Approximate and Plausible Reasoning with Applications to Expert Systems," *IEEE Transactions, PAMI*, **PAMI-7**, 3, pp. 260–283 (1985).

(4) Ishizuka, M., "Use of Uncertain Knowledge," *Keisoku to Seigyo*," **22**, 9, pp. 774–779 (1983) (in Japanese).

(5) Ishizuka, M., Fu, K. S., and Yao, J. T. P., "Rule-Based Damage Assessment System for Existing Structures," *Solid Mechanics Archives*, **8**, pp. 99–118 (1983).

(6) Ishizuka, M., Fu, K. S., and Yao, J. T. P., "SPERIL: An Expert System for Damage Assessment of Existing Structures," Sixth International Conf. on Pattern Recognition, Munich (1982).

(7) Ishizuka, M., "An Expert System for Structural Damage Assessment," *Transactions of the Information Processing Society of Japan*, **24**, 3, pp. 357–363 (1983) (in Japanese).

(8) Ogawa, H., Fu, K. S., and Yao, J. T. P., "SPERIL-II: An Expert System for Damage Assessment of Existing Structures," in Gupta, M. M., ed., *Approximate Reasoning in Expert Systems*, Elsevier North-Holland Science Publishers, (1985).

(9) Ogawa, H., Fu, K. S., and Yao, J. T. P., "An Expert System for Damage Assessment of Existing Structures," First Conference on Artificial Intelligence Applications, Denver (1984).

(10) Ogawa, H., Fu, K. S., and Yao, J. T. P., "Knowledge Representation and Inference Control of SPERIL-II," ACM'84 Annual Conference: The Fifth Generation Challenge (1984).

(11) Yao, J. T. P., "Damage Assessment and Reliability Evaluation of Existing Structures," *Journal of Engineering Structures*, 1, pp. 245–251 (1979).

(12) Yao, J. T. P., *Safety and Reliability of Existing Structures*, Pitman Publishing Ltd. (1985).

(13) Shortliffe, E. H., *et al.*, "Knowledge Engineering for Medical Decision Making: A Review of Computer-Based Clinical Decision Aids," *Proceedings of IEEE*, **67**, 7, pp. 1207–1224 (1979).

(14) Kulikowski, C. A., "Artificial Intelligence Methods and Systems for Medical Consultation," *IEEE Transactions, PAMI*, **PAMI-2**, 2, pp. 464–476 (1980).

(15) Ogawa, H., Fu, K. S., and Yao, J. T. P., "Inference Procedures under Uncertainty for the Problem-Reduction Method," *Information Science*, **28**, pp. 179–206 (1982).

(16) Ishizuka, M., "Inference Methods Based on Extended Dempster & Shafer's Theory for Problems with Uncertainty/Fuzziness," *New Generation Computing*, **1**, 2, pp. 159–168 (1983).

(17) Ogawa, H., Fu, K. S., and Yao, J. T. P., "A Rule-Based Inference with Fuzzy Set for Structural Damage Assessment," in Gupta, M. M., ed., *Approximate Reasoning in Decision Analysis*, North-Holland (1982).

(18) Dempster, A. P., "Upper and Lower Probabilities Induced by a Multivalued Mapping," *Annals of Mathematical Statistics*, **38**, pp. 325–339 (1967).

(19) Shafer, G., *A Mathematical Theory of Evidence*, Princeton University Press, Princeton, N. J. (1976).

(20) Ishizuka, M., Dempster & Shafer's Probability Theory, *Denshi Tsushin Gakkaishi* (Journal IECE of Japan), **66**, 9, pp. 900–903 (1983) (in Japanese).

(21) Shortliffe, E. H., *Computer-Based Medical Consultation: MYCIN*, American Elsevier, New York (1976).

(22) Duda, R. O., Hart, P., and Nilsson, N. J., *Subjective Baysian Methods for Rule-Based Inference Systems*, NCC (1976).

(23) Garvey, T. D., et al., "An Inference Technique for Integrating Knowledge from Disparate Sources," Seventh International Conference on Artificial Intelligence (1981).

(24) Friedman, L., "Extended Plausible Inference," *ibid.* (1981).

(25) Barnett, J. A., "Computational Method for a Mathematical Theory of Evidence," *ibid.* (1981).

(26) Tsuboi, K., Ishizuka, M., and Takanashi, K., "An Interactive Expert System for Structural Damage Assessment with Images and Graphics," National Convention Record of Institute of Image Electronics Engineers of Japan, **8** (1986) (in Japanese).

(27) J. Yen, "Generalizing the Dempster–Shafer Theory to Fuzzy Sets," *IEEE Transactions on SMC*, Vol. 20, No. 3, pp. 559–570 (1990).

INDEX

265

ISBN 0-12-685245-6